Pandas Cookbook
Third Edition

Practical recipes for scientific computing, time series, and exploratory data analysis using Python

William Ayd

Matthew Harrison

Pandas Cookbook

Third Edition

Senior Publishing Product Manager: Tushar Gupta
Acquisition Editor – Peer Reviews: Jane Dsouza
Project Editor: Janice Gonsalves
Content Development Editor: Shazeen Iqbal
Copy Editor: Safis Editing
Technical Editor: Gaurav Gavas
Proofreader: Safis Editing
Indexer: Manju Arasan
Presentation Designer: Ajay Patule
Developer Relations Marketing Executive: Vignesh Raju

First published: October 2017
Second edition: February 2020
Third edition: October 2024

Production reference: 2030725

Published by Packt Publishing Ltd.
Grosvenor House
11 St Paul's Square
Birmingham
B3 1RB, UK.

ISBN 978-1-83620-587-6

www.packt.com

Foreword

I started building what became the pandas project in 2008, during a rather different era in statistical computing and what we now call data science. At that time, data analysis was commonly performed using databases and SQL, Microsoft Excel, proprietary programming environments, and open-source languages like R. Python had a small but growing scientific computing community, yet it had little traction in statistical analysis and business analytics. While pandas began as my personal toolbox for data analysis work in Python, after a few years, it took on a life of its own as it became clear to me that Python had the potential to become a mainstream language for data analysis using open-source software.

Until 2013, I actively developed and maintained pandas for about five years. Since then, it has been community-maintained by a small, passionate group of core developers and thousands of community contributors. I published one of the first books to teach users about pandas in 2012, but today there are many books and online resources catering to different audiences. Some books focus mainly on explaining how to use the features of pandas, while others use pandas as an essential data manipulation tool as part of learning how to do data science or machine learning.

I've known Will Ayd and Matt Harrison for many years and have admired the work that they have done as open-source developers and educators for the Python community. Will is a member of the pandas core team and has built and maintained many of the features that are discussed in this book. Matt is an author of many successful Python books and possesses an amazing track record as a trainer and educator of Python programming, pandas, and other data science tools. This is a trustworthy duo to teach you how to do things the right way.

I am excited to see the third edition of this book come together. It is an excellent resource full of practical solutions to problems you will encounter in your data analysis work in Python. It covers the essential features of pandas while delving into more advanced functionality and features that were only added to the library in the last few years.

Wes McKinney
Creator of the pandas and Ibis projects
Co-creator of Apache Arrow

Contributors

About the authors

William Ayd is a core maintainer of the pandas project, serving in that role since 2018. For over a decade working as a consultant, Will has helped countless clients get the most value from their data using pandas and the open-source ecosystem surrounding it.

Matt Harrison has been using Python since 2000. He runs MetaSnake, which provides corporate training for Python and data science. He is the author of *Machine Learning Pocket Reference*, the best-selling *Illustrated Guide to Python 3*, and *Learning the Pandas Library*, among other books.

About the reviewer

Simon Hawkins holds a master's degree in aeronautical engineering from Imperial College, London. Early in his career, he worked exclusively in the defense and nuclear sectors as a technology analyst, specializing in modeling capabilities and simulation techniques for high-integrity equipment. He later transitioned into e-commerce, where his focus shifted to data analysis. Today, Simon is passionate about data science and is an active member of the pandas core development team.

Join our community on Discord

Join our community's Discord space for discussions with the authors and other readers:

https://packt.link/pandas

Table of Contents

Preface .... **xiii**

Making the Most Out of This Book – Get to Know Your Free Benefits xix

Chapter 1: pandas Foundations **1**

Importing pandas 2

Series 2

DataFrame 4

Index 5

Series attributes 7

DataFrame attributes 8

Chapter 2: Selection and Assignment **11**

Basic selection from a Series 12

Basic selection from a DataFrame 16

Position-based selection of a Series 17

Position-based selection of a DataFrame 19

Label-based selection from a Series 21

Label-based selection from a DataFrame 24

Mixing position-based and label-based selection 25

DataFrame.filter 28

Selection by data type 29

Selection/filtering via Boolean arrays 30

Selection with a MultiIndex – A single level 33

Selection with a MultiIndex – Multiple levels 34

Selection with a MultiIndex – a DataFrame 37

Item assignment with .loc and .iloc 38

DataFrame column assignment 39

Chapter 3: Data Types 45

Integral types 46

Floating point types 48

Boolean types 50

String types 51

Missing value handling 53

Categorical types 56

Temporal types – datetime 60

Temporal types – timedelta 65

Temporal PyArrow types 67

PyArrow List types 68

PyArrow decimal types 70

NumPy type system, the object type, and pitfalls 72

Chapter 4: The pandas I/O System 79

CSV – basic reading/writing 80

CSV – strategies for reading large files 85

Microsoft Excel – basic reading/writing 93

Microsoft Excel – finding tables in non-default locations 95

Microsoft Excel – hierarchical data 97

SQL using SQLAlchemy 99

SQL using ADBC 101

Apache Parquet 104

JSON 108

HTML 116

Pickle 119

Third-party I/O libraries 121

Chapter 5: Algorithms and How to Apply Them 123

Basic pd.Series arithmetic 124

Basic pd.DataFrame arithmetic 129

Aggregations 132

Transformations 136

Map 137

Apply 140

Summary statistics 142

Binning algorithms 143

One-hot encoding with pd.get_dummies 148

Chaining with .pipe 149

Selecting the lowest-budget movies from the top 100 152

Calculating a trailing stop order price 155

Finding the baseball players best at... 159

Understanding which position scores the most per tea 161

Chapter 6: Visualization   165

Creating charts from aggregated data 166

Plotting distributions of non-aggregated data 182

Further plot customization with Matplotlib 189

Exploring scatter plots 193

Exploring categorical data 202

Exploring continuous data 209

Using seaborn for advanced plots 214

Chapter 7: Reshaping DataFrames   225

Concatenating pd.DataFrame objects 226

Merging DataFrames with pd.merge 231

Joining DataFrames with pd.DataFrame.join 240

Reshaping with pd.DataFrame.stack and pd.DataFrame.unstack 242

Reshaping with pd.DataFrame.melt 247

Reshaping with pd.wide_to_long 250

Reshaping with pd.DataFrame.pivot and pd.pivot_table 252

Reshaping with pd.DataFrame.explode 257

Transposing with pd.DataFrame.T 261

Join our community on Discord 263

Chapter 8: Group By   265

Group by basics 266

Grouping and calculating multiple columns 270

Group by apply 275

Window operations 278

Selecting the highest rated movies by year 286

Comparing the best hitter in baseball across years 290

Chapter 9: Temporal Data Types and Algorithms   **299**

Timezone handling 300

DateOffsets 303

Datetime selection 307

Resampling 310

Aggregating weekly crime and traffic accidents 315

Calculating year-over-year changes in crime by category 318

Accurately measuring sensor-collected events with missing values 322

Chapter 10: General Usage and Performance Tips   **333**

Avoid dtype=object 333

Be cognizant of data sizes 338

Use vectorized functions instead of loops 340

Avoid mutating data 342

Dictionary-encode low cardinality data 343

Test-driven development features 343

Chapter 11: The pandas Ecosystem   **349**

Foundational libraries 350

   NumPy • 350

   PyArrow • 351

Exploratory data analysis 352

   YData Profiling • 352

Data validation 355

   Great Expectations • 356

Visualization 359

   Plotly • 360

   PyGWalker • 361

Data science 362

   scikit-learn • 362

   XGBoost • 364

Databases 365

   DuckDB • 365

Other DataFrame libraries 367

Ibis • 367

Dask • 370

Polars • 370

cuDF • 372

Other BooksYou May Enjoy **375**

Index **379**

Preface

pandas is a library for creating and manipulating structured data with Python. What do I mean by structured? I mean tabular data in rows and columns like what you would find in a spreadsheet or database. Data scientists, analysts, programmers, engineers, and others are leveraging it to mold their data.

pandas is limited to "small data" (data that can fit in memory on a single machine). However, the syntax and operations have been adopted by or inspired other projects: PySpark, Dask, and cuDF, among others. These projects have different goals, but some of them will scale out to big data. So, there is value in understanding how pandas works as the features are becoming the de facto API for interacting with structured data.

I, Will Ayd, have been a core maintainer of the pandas library since 2018. During that time, I have had the pleasure of contributing to and collaborating on a host of other open source projects in the same ecosystem, including but not limited to Arrow, NumPy and Cython.

I also consult for a living, utilizing the same ecosystem that I contribute to. Using the best open source tooling, I help clients develop data strategies, implement processes and patterns, and train associates to stay ahead of the ever-changing analytics curve. I strongly believe in the freedom that open source tooling provides, and have proven that value to many companies.

If your company is interested in optimizing your data strategy, feel free to reach out (will_ayd@ innobi.io).

Who this book is for

This book contains a huge number of recipes, ranging from very simple to advanced. All recipes strive to be written in clear, concise, and modern idiomatic pandas code. The *How it works* sections contain extremely detailed descriptions of the intricacies of each step of the recipe. Often, in the *There's more...* section, you will get what may seem like an entirely new recipe. This book is densely packed with an extraordinary amount of pandas code.

While not strictly required, users are advised to read the book chronologically. The recipes are structured in such a way that they first introduce concepts and features using very small, directed examples, but continuously build from there into more complex applications.

Due to the wide range of complexity, this book can be useful to both novice and everyday users alike. It has been my experience that even those who use pandas regularly will not master it without being exposed to idiomatic pandas code. This is somewhat fostered by the breadth that pandas offers. There are almost always multiple ways of completing the same operation, which can have users get the result they want but in a very inefficient manner. It is not uncommon to see an order of magnitude or more in performance difference between two sets of pandas solutions to the same problem.

The only real prerequisite for this book is a fundamental knowledge of Python. It is assumed that the reader is familiar with all the common built-in data containers in Python, such as lists, sets, dictionaries, and tuples.

What this book covers

Chapter 1, *pandas Foundations*, introduces the main pandas objects, namely, `Series`, DataFrames, and `Index`.

Chapter 2, *Selection and Assignment*, shows you how to sift through the data that you have loaded into any of the pandas data structures.

Chapter 3, *Data Types*, explores the type system underlying pandas. This is an area that has evolved rapidly and will continue to do so, so knowing the types and what distinguishes them is invaluable information.

Chapter 4, *The pandas I/O System*, shows why pandas has long been a popular tool to read from and write to a variety of storage formats.

Chapter 5, *Algorithms and How to Apply Them*, introduces you to the foundation of performing calculations with the pandas data structures.

Chapter 6, *Visualization*, shows you how pandas can be used directly for plotting, alongside the seaborn library which integrates well with pandas.

Chapter 7, *Reshaping DataFrames*, discusses the many ways in which data can be transformed and summarized robustly via the pandas `pd.DataFrame`.

Chapter 8, *Group By*, showcases how to segment and summarize subsets of your data contained within a `pd.DataFrame`.

Chapter 9, *Temporal Data Types and Algorithms*, introduces users to the date/time types which underlie time-series-based analyses that pandas is famous for and highlights usage against real data.

Chapter 10, *General Usage/Performance Tips*, goes over common pitfalls users run into when using pandas, and showcases the idiomatic solutions.

Chapter 11, *The pandas Ecosystem*, discusses other open source libraries that integrate, extend, and/ or complement pandas.

To get the most out of this book

There are a couple of things you can do to get the most out of this book. First, and most importantly, you should download all the code, which is stored in Jupyter Notebook. While reading through each recipe, run each step of code in the notebook. Make sure you explore on your own as you run through the code. Second, have the pandas official documentation open (http://pandas.pydata.org/pandas-docs/stable/) in one of your browser tabs. The pandas documentation is an excellent resource containing over 1,000 pages of material. There are examples for most of the pandas operations in the documentation, and they will often be directly linked from the *See also* section. While it covers the basics of most operations, it does so with trivial examples and fake data that don't reflect situations that you are likely to encounter when analyzing datasets from the real world.

What you need for this book

pandas is a third-party package for the Python programming language and, as of the printing of this book, is transitioning from the 2.x to the 3.x series. The examples in this book should work with a minimum pandas version of 2.0 along with Python versions 3.9 and above.

The code in this book will make use of the pandas, NumPy, and PyArrow libraries. Jupyter Notebook files are also a popular way to visualize and inspect code. All of these libraries should be installable via pip or the package manager of your choice. For pip users, you can run:

```
python -m pip install pandas numpy pyarrow notebook
```

Download the example code files

You can download the example code files for this book from your account at www.packt.com. If you purchased this book elsewhere, you can visit www.packtpub.com/support/errata and register to have the files emailed directly to you.

You can download the code files by following these steps:

1. Log in or register at www.packt.com.
2. Select the **Support** tab.
3. Click on **Code Downloads**.
4. Enter the name of the book in the **Search** box and follow the on-screen instructions.

The code bundle for the book is also hosted on GitHub at https://github.com/WillAyd/Pandas-Cookbook-Third-Edition. In case there is an update to the code, it will be updated in the existing GitHub repository.

Running a Jupyter notebook

The suggested method to work through the content of this book is to have a Jupyter notebook up and running so that you can run the code while reading through the recipes. Following along on your computer allows you to go off exploring on your own and gain a deeper understanding than by just reading the book alone.

After installing Jupyter notebook, open a Command Prompt (type cmd at the search bar on Windows, or open Terminal on Mac or Linux) and type:

```
jupyter notebook
```

It is not necessary to run this command from your home directory. You can run it from any location, and the contents in the browser will reflect that location. Although we have now started the Jupyter Notebook program, we haven't actually launched a single individual notebook where we can start developing in Python. To do so, you can click on the **New** button on the right-hand side of the page, which will drop down a list of all the possible kernels available for you to use. If you are working from a fresh installation, then you will only have a single kernel available to you (Python 3). After selecting the Python 3 kernel, a new tab will open in the browser, where you can start writing Python code.

You can, of course, open previously created notebooks instead of beginning a new one. To do so, navigate through the filesystem provided in the Jupyter Notebook browser home page and select the notebook you want to open. All Jupyter Notebook files end in .ipynb.

Alternatively, you may use cloud providers for a notebook environment. Both Google and Microsoft provide free notebook environments that come preloaded with pandas.

Download the color images

We also provide a PDF file that has color images of the screenshots/diagrams used in this book. You can download it here: https://packt.link/gbp/9781836205876.

Conventions

There are a number of text conventions used throughout this book.

CodeInText: Indicates code words in text, database table names, folder names, filenames, file extensions, pathnames, dummy URLs, user input, and Twitter/X handles. Here is an example: "You may need to install xlwt or openpyxl to write XLS or XLSX files, respectively."

A block of code is set as follows:

```
import pandas as pd
import numpy as np
movies = pd.read_csv("data/movie.csv")
movies
```

Bold: Indicates an important word, or words that you see on the screen. Here is an example: "**Select System info** from the **Administration** panel."

Italics: Indicates *terminology* that has extra importance within the context of the writing.

Important notes

Appear like this.

Tips

Appear like this.

Assumptions for every recipe

It should be assumed that at the beginning of each recipe, pandas, NumPy, PyArrow, and Matplotlib are imported into the namespace:

```
import numpy as np
import pyarrow as pa
import pandas as pd
```

Dataset descriptions

There are about two dozen datasets that are used throughout this book. It can be very helpful to have background information on each dataset as you complete the steps in the recipes. A detailed description of each dataset may be found in the `dataset_descriptions` Jupyter Notebook file found at `https://github.com/WillAyd/Pandas-Cookbook-Third-Edition`. For each dataset, there will be a list of the columns, information about each column, and notes on how the data was procured.

Sections

In this book, you will find several headings that appear frequently.

To give clear instructions on how to complete a recipe, we may use some or all of the following sections:

How to do it

This section contains the steps required to follow the recipe.

How it works

This section usually consists of a detailed explanation of what happened in the previous section.

There's more...

This section consists of additional information about the recipe in order to make you more knowledgeable about the recipe.

Get in touch

Feedback from our readers is always welcome.

General feedback: If you have questions about any aspect of this book, mention the book title in the subject of your message and email us at `customercare@packtpub.com`.

Errata: Although we have taken every care to ensure the accuracy of our content, mistakes do happen. If you have found a mistake in this book, we would be grateful if you would report this to us. Please visit, www.packtpub.com/support/errata, selecting your book, clicking on the **Errata Submission Form** link, and entering the details.

Piracy: If you come across any illegal copies of our works in any form on the internet, we would be grateful if you would provide us with the location address or website name. Please contact us at copyright@packt.com with a link to the material.

If you are interested in becoming an author: If there is a topic that you have expertise in and you are interested in either writing or contributing to a book, please visit authors.packtpub.com.

Reviews

Please leave a review. Once you have read and used this book, why not leave a review on the site that you purchased it from? Potential readers can then see and use your unbiased opinion to make purchase decisions, we at Packt can understand what you think about our products, and our authors can see your feedback on their book. Thank you!

For more information about Packt, please visit packt.com.

Leave a Review!

Thank you for purchasing this book from Packt Publishing—we hope you enjoy it! Your feedback is invaluable and helps us improve and grow. Once you've completed reading it, please take a moment to leave an Amazon review; it will only take a minute, but it makes a big difference for readers like you.

https://packt.link/r/1836205872

Making the Most Out of This Book — Get to Know Your Free Benefits

Unlock exclusive free benefits that come with your purchase, thoughtfully crafted to super-charge your learning journey and help you learn without limits.

UNLOCK NOW

Note: Have your purchase invoice ready before you begin.
https://www.packtpub.com/unlock/9781836205876

Figure 1.1: Next-Gen Reader, AI Assistant (Beta), and Free PDF access

Enhanced reading experience with our Next-gen Reader:

　Multi-device progress sync: Learn from any device with seamless progress sync.

　Highlighting and Notetaking: Turn your reading into lasting knowledge.

　Bookmarking: Revisit your most important learnings anytime.

　Dark mode: Focus with minimal eye strain by switching to dark or sepia modes.

Learn smarter using our AI assistant (Beta):

　Summarize it: Summarize key sections or an entire chapter.

　AI code explainers: In Packt Reader, click the "Explain" button above each code block for AI-powered code explanations.

Note: AI Assistant is part of next-gen Packt Reader and is still in beta.

Learn anytime, anywhere:

　Access your content offline with DRM-free PDF and ePub versions—compatible with your favorite e-readers.

Unlock Your Book's Exclusive Benefits

Your copy of this book comes with the following exclusive benefits:

- Next-gen Packt Reader
- AI assistant (beta)
- DRM-free PDF/ePub downloads

Use the following guide to unlock them if you haven't already. The process takes just a few minutes and needs to be done only once.

How to unlock these benefits in three easy steps

Step 1

Have your purchase invoice for this book ready, as you'll need it in *Step 3*. If you received a physical invoice, scan it on your phone and have it ready as either a PDF, JPG, or PNG.

For more help on finding your invoice, visit `https://www.packtpub.com/unlock-benefits/help`.

 Note: Bought this book directly from Packt? You don't need an invoice. After completing Step 2, you can jump straight to your exclusive content.

Step 2

Scan the following QR code or visit `https://www.packtpub.com/unlock/9781836205876`:

Step 3

Sign in to your Packt account or create a new one for free. Once you're logged in, upload your invoice. It can be in PDF, PNG, or JPG format and must be no larger than 10 MB. Follow the rest of the instructions on the screen to complete the process.

Need help?

If you get stuck and need help, visit `https://www.packtpub.com/unlock-benefits/help` for a detailed FAQ on how to find your invoices and more. The following QR code will take you to the help page directly:

 Note: If you are still facing issues, reach out to `customercare@packt.com`.

1

pandas Foundations

The **pandas** library is useful for dealing with structured data. What is structured data? Data that is stored in tables, such as CSV files, Excel spreadsheets, or database tables, is all structured. Unstructured data consists of free-form text, images, sound, or video. If you find yourself dealing with structured data, pandas will be of great utility to you.

`pd.Series` is a one-dimensional collection of data. If you are coming from Excel, you can think of this as a column. The main difference is that, like a column in a database, all of the values within `pd.Series` must have a single, homogeneous type.

`pd.DataFrame` is a two-dimensional object. Much like an Excel sheet or database table can be thought of as a collection of columns, `pd.DataFrame` can be thought of as a collection of `pd.Series` objects. Each `pd.Series` has a homogeneous data type, but the `pd.DataFrame` is allowed to be heterogeneous and store a variety of `pd.Series` objects with different data types.

`pd.Index` does not have a direct analogy with other tools. Excel may offer the closest with auto-numbered rows on the left-hand side of a worksheet, but those numbers tend to be for display purposes only. `pd.Index`, as you will find over the course of this book, can be used for selecting values, joining tables, and much more.

The recipes in this chapter will show you how to manually construct `pd.Series` and `pd.DataFrame` objects, customize the `pd.Index` object(s) associated with each, and showcase common attributes of the `pd.Series` and `pd.DataFrame` that you may need to inspect during your analyses.

We are going to cover the following recipes in this chapter:

- Importing pandas
- Series
- DataFrame
- Index
- Series attributes
- DataFrame attributes

Importing pandas

Most users of the **pandas** library will use an import alias so they can refer to it as pd. In general, in this book, we will not show the pandas and **NumPy** imports, but they look like this:

```
import pandas as pd
import numpy as np
```

💡 **Quick tip:** Enhance your coding experience with the **AI Code Explainer** and **Quick Copy** features. Open this book in the next-gen Packt Reader. Click the **Copy** button (1) to quickly copy code into your coding environment, or click the **Explain** button (2) to get the AI assistant to explain a block of code to you.

Copy   Explain
  1   2

```
function calculate(a, b) {
    return {sum: a + b};
};
```

🔒 **The next-gen Packt Reader** is included for free with the purchase of this book. Unlock it by scanning the QR code below or visiting https://www.packtpub.com/unlock/9781836205876.

While it is an optional dependency in the 2.x series of pandas, many examples in this book will also leverage the **PyArrow** library, which we assume to be imported as:

```
import pyarrow as pa
```

Series

The basic building block in pandas is a pd.Series, which is a one-dimensional array of data paired with a pd.Index. The index labels can be used as a simplistic way to look up values in the pd.Series, much like the Python dictionary built into the language uses key/value pairs (we will expand on this and much more pd.Index functionality in *Chapter 2, Selection and Assignment*).

The following section demonstrates a few ways of creating a pd.Series directly.

How to do it

The easiest way to construct a pd.Series is to provide a sequence of values, like a list of integers:

```
pd.Series([0, 1, 2])
```

```
0    0
1    1
2    2
dtype: int64
```

A **tuple** is another type of sequence, making it valid as an argument to the pd.Series constructor:

```
pd.Series((12.34, 56.78, 91.01))
```

```
0    12.34
1    56.78
2    91.01
dtype: float64
```

When generating sample data, you may often reach for the Python range function:

```
pd.Series(range(0, 7, 2))
```

```
0    0
1    2
2    4
3    6
dtype: int64
```

In all of the examples so far, pandas will try and infer a proper data type from its arguments for you. However, there are times when you will know more about the type and size of your data than can be inferred. Providing that information explicitly to pandas via the dtype= argument can be useful to save memory or ensure proper integration with other typed systems, like SQL databases.

To illustrate this, let's use a simple range argument to fill a pd.Series with a sequence of integers. When we did this before, the inferred data type was a 64-bit integer, but we, as developers, may know that we never expect to store larger values in this pd.Series and would be fine with only 8 bits of storage (if you do not know the difference between an 8-bit and 64-bit integer, that topic will be covered in *Chapter 3*, *Data Types*). Passing dtype="int8" to the pd.Series constructor will let pandas know we want to use the smaller data type:

```
pd.Series(range(3), dtype="int8")
```

```
0    0
1    1
2    2
dtype: int8
```

A pd.Series can also have a name attached to it, which can be specified via the name= argument (if not specified, the name defaults to None):

```
pd.Series(["apple", "banana", "orange"], name="fruit")
```

```
0       apple
1       banana
2       orange
Name: fruit, dtype: object
```

DataFrame

While pd.Series is the building block, pd.DataFrame is the main object that comes to mind for users of pandas. pd.DataFrame is the primary and most commonly used object in pandas, and when people think of pandas, they typically envision working with a pd.DataFrame.

In most analysis workflows, you will be importing your data from another source, but for now, we will show you how to construct a pd.DataFrame directly (input/output will be covered in *Chapter 4, The pandas I/O System*).

How to do it

The most basic construction of a pd.DataFrame happens with a two-dimensional sequence, like a list of lists:

```
pd.DataFrame([
    [0, 1, 2],
    [3, 4, 5],
    [6, 7, 8],
])
```

```
    0   1   2
0   0   1   2
1   3   4   5
2   6   7   8
```

With a list of lists, pandas will automatically number the row and column labels for you. Typically, users of pandas will at least provide labels for columns, as it makes indexing and selecting from a pd.DataFrame much more intuitive (see *Chapter 2, Selection and Assignment*, for an introduction to indexing and selecting). To label your columns when constructing a pd.DataFrame from a list of lists, you can provide a columns= argument to the constructor:

```
pd.DataFrame([
    [1, 2],
    [4, 8],
], columns=["col_a", "col_b"])
```

	col_a	col_b
0	1	2
1	4	8

Instead of using a list of lists, you could also provide a dictionary. The keys of the dictionary will be used as column labels, and the values of the dictionary will represent the values placed in that column of the pd.DataFrame:

```python
pd.DataFrame({
    "first_name": ["Jane", "John"],
    "last_name": ["Doe", "Smith"],
})
```

	first_name	last_name
0	Jane	Doe
1	John	Smith

In the above example, our dictionary values were lists of strings, but the pd.DataFrame does not strictly require lists. Any sequence will work, including a pd.Series:

```python
ser1 = pd.Series(range(3), dtype="int8", name="int8_col")
ser2 = pd.Series(range(3), dtype="int16", name="int16_col")
pd.DataFrame({ser1.name: ser1, ser2.name: ser2})
```

	int8_col	int16_col
0	0	0
1	1	1
2	2	2

Index

When constructing both the pd.Series and pd.DataFrame objects in the previous sections, you likely noticed the values to the left of these objects starting at 0 and incrementing by 1 for each new row of data. The object responsible for those values is the pd.Index, highlighted in the following image:

```
pd.Series(["x", "y", "z"])

0    x
1    y
2    z
dtype: object
```

Figure 1.1: Default pd.Index, highlighted in red

In the case of a pd.DataFrame, you have a pd.Index not only to the left of the object (often referred to as the *row index* or even just *index*) but also above (often referred to as the *column index* or *columns*):

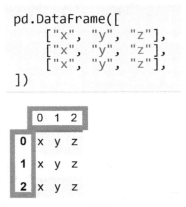

Figure 1.2: A pd.DataFrame with a row and column index

Unless explicitly provided, pandas will create an auto-numbered pd.Index for you (technically, this is a pd.RangeIndex, a subclass of the pd.Index class). However, it is very rare to use pd.RangeIndex for your columns, as referring to a column named City or Date is more expressive than referring to a column in the nth position. The pd.RangeIndex appears more commonly in the row index, although you may still want custom labels to appear there as well. More advanced selection operations with the default pd.RangeIndex and custom pd.Index values will be covered in *Chapter 2, Selection and Assignment*, to help you understand different use cases, but for now, let's just look at how you would override the construction of the row and column pd.Index objects during pd.Series and pd.DataFrame construction.

How to do it

When constructing a pd.Series, the easiest way to change the row index is by providing a sequence of labels to the index= argument. In this example, the labels dog, cat, and human will be used instead of the default pd.RangeIndex numbered from 0 to 2:

```
pd.Series([4, 4, 2], index=["dog", "cat", "human"])
```

```
dog       4
cat       4
human     2
dtype: int64
```

If you want finer control, you may want to construct the pd.Index yourself before passing it as an argument to index=. In the following example, the pd.Index is given the name animal, and the pd.Series itself is named num_legs, providing more context to the data:

```
index = pd.Index(["dog", "cat", "human"], name="animal")
pd.Series([4, 4, 2], name="num_legs", index=index)
```

```
animal
dog           4
cat           4
human         2
Name: num_legs, dtype: int64
```

A pd.DataFrame uses a pd.Index for both dimensions. Much like with the pd.Series constructor, the index= argument can be used to specify the row labels, but you now also have the columns= argument to control the column labels:

```
pd.DataFrame([
    [24, 180],
    [42, 166],
], columns=["age", "height_cm"], index=["Jack", "Jill"])
```

```
        age    height_cm
Jack    24     180
Jill    42     166
```

Series attributes

Once you have a pd.Series, there are quite a few attributes you may want to inspect. The most basic attributes can tell you the type and size of your data, which is often the first thing you will inspect when reading in data from a data source.

How to do it

Let's start by creating a pd.Series that has a name, alongside a custom pd.Index, which itself has a name. Although not all of these elements are required, having them will help us more clearly understand what the attributes we access through this recipe are actually showing us:

```
index = pd.Index(["dog", "cat", "human"], name="animal")
ser = pd.Series([4, 4, 2], name="num_legs", index=index)
ser
```

```
animal
dog       4
cat       4
human     2
Name: num_legs, dtype: int64
```

The first thing users typically want to know about their data is the type of pd.Series. This can be inspected via the pd.Series.dtype attribute:

```
ser.dtype
```

```
dtype('int64')
```

The name may be inspected via the pd.Series.name attribute. The data we constructed in this recipe was created with the name="num_legs" argument, which is what you will see when accessing this attribute (if not provided, this will return None):

```
ser.name
```

```
num_legs
```

The associated pd.Index can be accessed via pd.Series.index:

```
ser.index
```

```
Index(['dog', 'cat', 'human'], dtype='object', name='animal')
```

The name of the associated pd.Index can be accessed via pd.Series.index.name:

```
ser.index.name
```

```
animal
```

The shape can be accessed via pd.Series.shape. For a one-dimensional pd.Series, the shape is returned as a one-tuple where the first element represents the number of rows:

```
ser.shape
```

```
3
```

The size (number of elements) can be accessed via pd.Series.size:

```
ser.size
```

```
3
```

The Python built-in function len can show you the length (number of rows):

```
len(ser)
```

```
3
```

DataFrame attributes

The pd.DataFrame shares many of the attributes of the pd.Series, with some slight differences. Generally, pandas tries to share as many attributes as possible between the pd.Series and pd.DataFrame, but the two-dimensional nature of the pd.DataFrame makes it more natural to express some things in plural form (for example, the .dtype attribute becomes .dtypes) and gives us a few more attributes to inspect (for example, .columns exists for a pd.DataFrame but not for a pd.Series).

How to do it

Much like we did in the previous section, we are going to construct a pd.DataFrame with a custom pd.Index in the rows, while also using custom labels in the columns. This will be more helpful when inspecting the various attributes:

```
index = pd.Index(["Jack", "Jill"], name="person")
df = pd.DataFrame([
    [24, 180, "red"],
    [42, 166, "blue"],
], columns=["age", "height_cm", "favorite_color"], index=index)
df
```

```
            age     height_cm     favorite_color
person
Jack        24      180           red
Jill        42      166           blue
```

The types of each column can be inspected via the pd.DataFrame.dtypes attribute. This attribute returns a pd.Series where each row shows the data type corresponding to each column in our pd.DataFrame:

```
df.dtypes
```

```
age                int64
height_cm          int64
favorite_color     object
dtype: object
```

The *row index* can be accessed via pd.DataFrame.index:

```
df.index
```

```
Index(['Jack', 'Jill'], dtype='object', name='person')
```

The *column index* can be accessed via pd.DataFrame.columns:

```
df.columns
```

```
Index(['age', 'height_cm', 'favorite_color'], dtype='object')
```

The shape can be accessed via pd.DataFrame.shape. For a two-dimensional pd.DataFrame, the shape is returned as a two-tuple where the first element represents the number of rows and the second element represents the number of columns:

```
df.shape
```

```
2     3
```

The size (number of elements) can be accessed via pd.DataFrame.size:

```
df.size
```

```
6
```

The Python built-in function len can show you the length (number of rows):

```
len(df)
```

```
2
```

Join our community on Discord

Join our community's Discord space for discussions with the authors and other readers:

```
https://packt.link/pandas
```

2

Selection and Assignment

In the previous chapter, we looked at how to create a `pd.Series` and `pd.DataFrame`, and we also looked at their relationship to the `pd.Index`. With a foundation in *constructors*, we now shift focus to the crucial processes of *selection* and *assignment*. Selection, also referred to as *indexing*, is considered a *getter*; i.e., it is used to retrieve values from a pandas object. Assignment, by contrast, is a *setter* that is used to update values.

The recipes in this chapter start out by showing you how to retrieve values from `pd.Series` and `pd.DataFrame` objects, with ever-increasing complexity. We will eventually introduce the `pd.MultiIndex`, which can be used to select data hierarchically, before finally ending with an introduction to the assignment operators. The pandas API takes great care to reuse many of the same methods for selection and assignment, which ultimately allows you to be very expressive in how you would like to interact with your data.

By the end of this chapter, you will be adept at efficiently retrieving data from and updating values within your pandas objects. We are going to cover the following recipes in this chapter:

- Basic selection from a Series
- Basic selection from a DataFrame
- Position-based selection of a Series
- Position-based selection of a DataFrame
- Label-based selection from a Series
- Label-based selection from a DataFrame
- Mixing position-based and label-based selection
- DataFrame.filter
- Selection by data type
- Selection/filtering via Boolean arrays
- Selection with a MultiIndex – A single level
- Selection with a MultiIndex – Multiple levels
- Selection with a MultiIndex – a DataFrame

- Item assignment with .loc and .iloc
- DataFrame column assignment

Basic selection from a Series

Selection from a pd.Series involves accessing elements either by their position or by their label. This is akin to accessing elements in a list by their index or in a dictionary by their key, respectively. The versatility of the pd.Series object allows intuitive and straightforward data retrieval, making it an essential tool for data manipulation.

The pd.Series is considered a *container* in Python, much like the built-in list, tuple, and dict objects. As such, for simple selection operations, the first place users turn to is the Python index operator, using the [ ] syntax.

How to do it

To introduce the basics of selection, let's start with a very simple pd.Series:

```
ser = pd.Series(list("abc") * 3)
ser
```

```
0    a
1    b
2    c
3    a
4    b
5    c
6    a
7    b
8    c
dtype: object
```

In Python, you've already discovered that the [ ] operator can be used to select elements from a *container*; i.e., some_dictionary[0] will give you the value associated with a key of 0. With a pd.Series, basic selection behaves similarly:

```
ser[3]
```

```
a
```

With the expression ser[3], pandas tries to find the label 3 in the index of the pd.Series and, assuming only one match, returns the value associated with that label.

Instead of selecting the associated value from the pd.Series, alternatively, you may want a pd.Series returned, as doing so helps you keep the label 3 associated with the data element "a." With pandas, you can do this by providing a list argument that contains a single element:

```
ser[[3]]
```

```
3    a
dtype: object
```

Expanding on the usage of a list argument, you can select multiple values from the pd.Series if your list contains multiple elements:

```
ser[[0, 2]]
```

```
0    a
2    c
dtype: object
```

Assuming you use the default index, you can use slice arguments that work very similarly to slicing a Python list. For example, to get up to (but not including) the element at position 3 of a pd.Series, you can use:

```
ser[:3]
```

```
0    a
1    b
2    c
dtype: object
```

Negative slice indexers are not a problem for pandas. The following code will select the last four elements of the pd.Series:

```
ser[-4:]
```

```
5    c
6    a
7    b
8    c
dtype: object
```

You can even provide slices with *start* and *stop* arguments. The following code will retrieve all elements of the pd.Series, starting in position 2 and up to (but not including) position 6:

```
ser[2:6]
```

```
2    c
3    a
4    b
5    c
dtype: object
```

This final example on slices uses *start*, *stop* and *step* arguments to grab every third element, starting at position 1 and stopping when position 8 is encountered:

```
ser[1:8:3]
```

```
1     b
4     b
7     b
dtype: object
```

Selection still works when providing your own pd.Index values. Let's create a small pd.Series with string index labels to illustrate:

```
ser = pd.Series(range(3), index=["Jack", "Jill", "Jayne"])
ser
```

```
Jack     0
Jill     1
Jayne    2
dtype: int64
```

Selection via ser["Jill"] will scan the index for the string Jill and return the corresponding element:

```
ser["Jill"]
```

```
1
```

Once again, providing a single-element list argument will ensure that you receive a pd.Series in return instead of a single value:

```
ser[["Jill"]]
```

```
Jill    1
dtype: int64
```

There's more...

A common pitfall when using the [] operator is to assume that selection with integer arguments works the same as when selecting from a Python list. This is *only* true when you use the default pd.Index, which is auto-numbered, starting at 0 (this is technically called a pd.RangeIndex).

When not using a pd.RangeIndex, extra attention must be paid to the behavior. To illustrate, let's start with a small pd.Series, which still uses integers in its pd.Index, but does not use an auto-incrementing sequence that starts at 0:

```
ser = pd.Series(list("abc"), index=[2, 42, 21])
ser
```

```
2     a
42    b
21    c
dtype: object
```

It is important to note that an integer argument selects by *label* and not by *position*; i.e., the following code will return the value associated with a label of 2, not the value in position 2:

```
ser[2]
```

```
a
```

While the integer argument matches by label and not by position, slicing still works positionally. The following example does not stop when encountering the number 2 and, instead, gives the first two elements back:

```
ser[:2]
```

```
2     a
42    b
dtype: object
```

Users should also be familiar with selection behavior when working with a non-unique pd.Index. Let's create a small pd.Series where the number 1 appears twice in our row index:

```
ser = pd.Series(["apple", "banana", "orange"], index=[0, 1, 1])
ser
```

```
0      apple
1      banana
1      orange
dtype: object
```

With this pd.Series, attempting to select the number 1 will *not* return a single value and, instead, return another pd.Series:

```
ser[1]
```

```
1      banana
1      orange
dtype: object
```

The fact that a selection like ser[1] can be thought to be done by position or label interchangeably when using the default pd.RangeIndex but, in actuality, selects by label with other pd.Index types can be the source of subtle bugs in user programs. Many users may *think* they are selecting the nth element, only to have that assumption break when their data changes. To resolve the ambiguity between wanting to select by *label* or by *position* with an integer argument, it is **highly recommended** to leverage the .loc and .iloc methods introduced later in this chapter.

Basic selection from a DataFrame

When using the [ ] operator with a pd.DataFrame, simple selection typically involves selecting data from the *column index* rather than the *row index*. This distinction is crucial for effective data manipulation and analysis. Columns in a pd.DataFrame can be accessed by their labels, making it easy to work with named data from a pd.Series within the larger pd.DataFrame structure.

Understanding this fundamental difference in selection behavior is key to utilizing the full power of a pd.DataFrame in pandas. By leveraging the [ ] operator, you can efficiently access and manipulate specific columns of data, setting the stage for more advanced operations and analyses.

How to do it

Let's start by creating a simple 3x3 pd.DataFrame. The values of the pd.DataFrame are not important, but we are intentionally going to provide our own column labels instead of having pandas create an auto-numbered column index for us:

```
df = pd.DataFrame(np.arange(9).reshape(3, -1), columns=["a", "b", "c"])
df
```

```
     a    b    c
0    0    1    2
1    3    4    5
2    6    7    8
```

To select a single column, use the [ ] operator with a scalar argument:

```
df["a"]
```

```
0    0
1    3
2    6
Name: a, dtype: int64
```

To select a single column but still get back a pd.DataFrame instead of a pd.Series, pass a single-element list:

```
df[["a"]]
```

```
     a
0    0
1    3
2    6
```

Multiple columns can be selected using a list:

```
df[["a", "b"]]
```

	a	b
0	0	1
1	3	4
2	6	7

In all of these examples, the arguments for [ ] have been selected from the columns, but providing a slice argument exhibits different behavior and will actually select from rows. Note that the following example selects all columns and the first two rows of data, not the other way around:

```
df[:2]
```

	a	b	c
0	0	1	2
1	3	4	5

There's more...

When using a list argument for the [ ] operator, you have the flexibility to specify the order of columns in the output. This allows you to customize the pd.DataFrame to suit your needs. The order of columns in the output will exactly match the order of labels provided as input. For example:

```
df[["a", "b"]]
```

	a	b
0	0	1
1	3	4
2	6	7

Swapping the order of the elements in the list as an argument to [ ] will swap the order of the columns in the resulting pd.DataFrame:

```
df[["b", "a"]]
```

	b	a
0	1	0
1	4	3
2	7	6

This feature is particularly useful when you need to reorder columns for presentation purposes, or when preparing data for export to CSV or Excel formats where a specific column order is required (see *Chapter 4, The pandas I/O System,* for more on the pandas I/O system).

Position-based selection of a Series

As discussed back in the *Basic selection from a DataFrame* section, using [ ] as a selection mechanism does not signal the clearest intent and can sometimes be downright confusing. The fact that ser[42] selects from a *label* matching the number 42 and not the 42nd row of a pd.Series is a common mistake for new users, and such an ambiguity can grow even more complex as you start trying to select two dimensions with the [ ] operator from a pd.DataFrame.

To clearly signal that you are trying to select by *position* instead of by *label*, you should use pd.Series.iloc.

How to do it

Let's create a pd.Series where we have an index using integral labels that are also non-unique:

```
ser = pd.Series(["apple", "banana", "orange"], index=[0, 1, 1])
ser
```

```
0      apple
1      banana
1      orange
dtype: object
```

To select a scalar, you can use pd.Series.iloc with an integer argument:

```
ser.iloc[1]
```

```
banana
```

Following the same patterns we have seen before, turning that integer argument into a list containing a single element will return a pd.Series instead of a scalar:

```
ser.iloc[[1]]
```

```
1      banana
dtype: object
```

Multiple integers in the list argument will select multiple elements of the pd.Series by position:

```
ser.iloc[[0, 2]]
```

```
0      apple
1      orange
dtype: object
```

Slices are a natural way of expressing a range of elements that you would select, and they pair very nicely as an argument to pd.Series.iloc:

```
ser.iloc[:2]
```

```
0      apple
1      banana
dtype: object
```

Position-based selection of a DataFrame

Much like with a pd.Series, integers, lists of integers, and slice objects are all valid arguments to DataFrame.iloc. However, with a pd.DataFrame, two arguments are required. The first argument handles selecting from the *rows*, and the second is responsible for the *columns*.

In most use cases, users reach for position-based selection when retrieving rows and label-based selection when retrieving columns. We will cover the latter in the *Label-based selection from a DataFrame* section and will show you how to combine both in the *Mixing position-based and label-based selection* section. However, when your row index uses the default pd.RangeIndex and the order of columns is significant, the techniques shown in this section will be of immense value.

How to do it

Let's create a pd.DataFrame with five rows and four columns:

```
df = pd.DataFrame(np.arange(20).reshape(5, -1), columns=list("abcd"))
df
```

```
   a   b   c   d
0  0   1   2   3
1  4   5   6   7
2  8   9   10  11
3  12  13  14  15
4  16  17  18  19
```

Passing two integer arguments to pd.DataFrame.iloc will return a scalar from that row and column position:

```
df.iloc[2, 2]
```

```
10
```

In some cases, you may not want to select individual values from a particular axis, opting instead for everything that axis has to offer. An empty slice object, :, allows you to do this; i.e., if you wanted to select *all* rows of data from the first column of a pd.DataFrame, you would use:

```
df.iloc[:, 0]
```

```
0     0
1     4
2     8
3     12
4     16
Name: a, dtype: int64
```

Flipping the order of the arguments to pd.DataFrame.iloc will change behavior. Instead of grabbing all rows for the first column, the below code selects all columns and only the first row of data:

```
df.iloc[0, :]
```

```
a    0
b    1
c    2
d    3
Name: 0, dtype: int64
```

Because the preceding examples only return one dimension of data, they implicitly attempt to *squeeze* the return value from a pd.DataFrame down to a pd.Series. Following the patterns we have seen many times already in this chapter, you can prevent that implicit dimensionality reduction by passing a single-element list argument for the axis, which is not an empty slice. For example, to select all rows for the first column but still get back a pd.DataFrame, you would opt for:

```
df.iloc[:, [0]]
```

```
     a
0    0
1    4
2    8
3    12
4    16
```

Reversing those arguments gives us the first row and all columns back as a pd.DataFrame:

```
df.iloc[[0], :]
```

```
     a    b    c    d
0    0    1    2    3
```

Lists can be used to select multiple elements from both the rows and columns. If we wanted the first and second rows paired with the last and second-to-last columns of our pd.DataFrame, you could opt for an expression like:

```
df.iloc[[0, 1], [-1, -2]]
```

```
     d    c
0    3    2
1    7    6
```

There's more...

Empty slices are valid arguments to .iloc. Both ser.iloc[:] and df.iloc[:, :] will return everything from each axis, essentially giving you a copy of the object.

Label-based selection from a Series

In pandas, `pd.Series.loc` is used to perform selection by label instead of by position. This method is particularly useful when you consider the `pd.Index` of your `pd.Series` to contain lookup values, much like the key in a Python dictionary, rather than giving importance to the order or position of data in your `pd.Series`.

How to do it

Let's create a `pd.Series` where we have a row index using integral labels that are also non-unique:

```
ser = pd.Series(["apple", "banana", "orange"], index=[0, 1, 1])
ser
```

```
0       apple
1       banana
1       orange
dtype: object
```

`pd.Series.loc` will select all rows where the index has a label of 1:

```
ser.loc[1]
```

```
1       banana
1       orange
dtype: object
```

Of course, you are not limited to integral labels in pandas. Let's see what this looks like with a `pd.Index` composed of string values:

```
ser = pd.Series([2, 2, 4], index=["dog", "cat", "human"], name="num_legs")
ser
```

```
dog       2
cat       2
human     4
Name: num_legs, dtype: int64
```

`pd.Series.loc` can select all rows where the index has a label of "dog":

```
ser.loc["dog"]
```

```
2
```

To select all rows where the index has a label of "dog" or "cat":

```
ser.loc[["dog", "cat"]]
```

```
dog     2
```

```
cat     2
Name: num_legs, dtype: int64
```

Finally, to select all rows up to and including the label `"cat"`:

```
ser.loc[:"cat"]
```

```
dog     2
cat     2
Name: num_legs, dtype: int64
```

There's more...

Understanding label-based selection with `pd.Series.loc` provides powerful capabilities to access and manipulate data in a `pd.Series`. While this method may seem straightforward, it offers nuances and behaviors that are important to grasp for effective data handling.

A very common mistake for users of all experience levels with pandas is to overlook the differences in behavior that slicing with `pd.Series.loc` has, compared to slicing in standard Python and the `pd.Series.iloc` case.

To walk through this, let's create a small Python list and a `pd.Series` with the same data:

```
values = ["Jack", "Jill", "Jayne"]
ser = pd.Series(values)
ser
```

```
0       Jack
1       Jill
2       Jayne
dtype: object
```

As you have already seen with lists and other containers built into the Python language, slicing returns values up to *but not including* the provided position:

```
values[:2]
```

```
Jack    Jill
```

Slicing with `pd.Series.iloc` matches this behavior, returning a `pd.Series` with the same exact length and elements as the Python list:

```
ser.iloc[:2]
```

```
0       Jack
1       Jill
dtype: object
```

But slicing with `pd.Series.loc` actually produces a different result:

```
ser.loc[:2]
```

```
0        Jack
1        Jill
2        Jayne
dtype: object
```

What is going on here? To try and get a grasp on this, it is important to remember that pd.Series.loc matches by label, not by position. The pandas library does something akin to a loop over each element in the pd.Series and its accompanying pd.Index, stopping at the point where it finds the value of 2 in the index. However, pandas cannot guarantee that there is only one value in the pd.Index with the value of 2, so it must continue going until it finds *something else*. You can see that in action if you try the same selection with a pd.Series that repeats the index label 2:

```
repeats_2 = pd.Series(range(5), index=[0, 1, 2, 2, 0])
repeats_2.loc[:2]
```

```
0        0
1        1
2        2
2        3
dtype: int64
```

This can seem downright devious if you expect your row index to contain integers, but the main use case for pd.Series.loc is for working with a pd.Index where position/ordering is not important (for that, use pd.Series.iloc). Taking string labels as a more practical example, the slicing behavior of pd.Series.loc becomes more natural. The following code can essentially be thought of as asking pandas to loop over the pd.Series until the label "xxx" is found in the row index, continuing until a new label is found:

```
ser = pd.Series(range(4), index=["zzz", "xxx", "xxx", "yyy"])
ser.loc[:"xxx"]
```

```
zzz      0
xxx      1
xxx      2
dtype: int64
```

In certain cases where you try to slice with pd.Series.loc but the index labels have no determinate ordering, pandas will end up raising an error:

```
ser = pd.Series(range(4), index=["zzz", "xxx", "yyy", "xxx"])
ser.loc[:"xxx"]
```

```
KeyError: "Cannot get right slice bound for non-unique label: 'xxx'"
```

Label-based selection from a DataFrame

As we discussed back in the *Position-based selection of a DataFrame* section, the most common use case with a pd.DataFrame is to use label-based selection when referring to columns and position-based selection when referring to rows. However, this is not an absolute requirement, and pandas allows you to use label-based selection from both the rows and columns.

When compared to other data analysis tools, the ability to select by label from the rows of a pd.DataFrame is a unique advantage to pandas. For users familiar with SQL, there is no real equivalent to this provided by the language; columns are very easy to select when placed in a SELECT clause, but rows can only be filtered via a WHERE clause. For users adept at Microsoft Excel, you could create two-dimensional structures using a pivot table, with both row labels and column labels, but your ability to select or refer to data within that pivot table is effectively limited.

For now, we will introduce selection for very small pd.DataFrame objects to get a feel for the syntax. In *Chapter 8, Reshaping Data Frames*, we will explore ways that you can create meaningful pd.DataFrame objects where row and column labels are significant. Combined with the knowledge introduced in this section, you will come to appreciate how unique this type of selection is to pandas, as well as how it can help you explore data in meaningful ways that other tools cannot express.

How to do it

Let's create a pd.DataFrame where we have indices composed of strings in both the rows and columns:

```
df = pd.DataFrame([
    [24, 180, "blue"],
    [42, 166, "brown"],
    [22, 160, "green"],
], columns=["age", "height_cm", "eye_color"], index=["Jack", "Jill", "Jayne"])
df
```

```
       age    height_cm     eye_color
Jack   24     180           blue
Jill   42     166           brown
Jayne  22     160           green
```

pd.DataFrame.loc can select by the row and column label:

```
df.loc["Jayne", "eye_color"]
```

```
green
```

To select all rows from the column with the label "age":

```
df.loc[:, "age"]
```

```
Jack      24
Jill      42
Jayne     22
Name: age, dtype: int64
```

To select all columns from the row with the label "Jack":

```
df.loc["Jack", :]
```

```
age              24
height_cm       180
eye_color      blue
Name: Jack, dtype: object
```

To select all rows from the column with the label "age", maintaining the pd.DataFrame shape:

```
df.loc[:, ["age"]]
```

```
        age
Jack     24
Jill     42
Jayne    22
```

To select all columns from the row with the label "Jack", maintaining the pd.DataFrame shape:

```
df.loc[["Jack"], :]
```

```
        age    height_cm    eye_color
Jack     24    180          blue
```

To select both rows and columns using lists of labels:

```
df.loc[["Jack", "Jill"], ["age", "eye_color"]]
```

```
        age    eye_color
Jack     24    blue
Jill     42    brown
```

Mixing position-based and label-based selection

Since pd.DataFrame.iloc is used for position-based selection and pd.DataFrame.loc is for label-based selection, users must take an extra step if attempting to select by label in one dimension and by position in another. As mentioned in previous sections, the majority of pd.DataFrame objects constructed will place heavy significance on the labels used for the columns, with little care for how those columns are ordered. The inverse is true for the rows, so being able to effectively mix and match both styles is of immense value.

How to do it

Let's start with a pd.DataFrame that uses the default auto-numbered pd.RangeIndex in the rows but has custom string labels for the columns:

```
df = pd.DataFrame([
    [24, 180, "blue"],
    [42, 166, "brown"],
    [22, 160, "green"],
], columns=["age", "height_cm", "eye_color"])
df
```

	age	height_cm	eye_color
0	24	180	blue
1	42	166	brown
2	22	160	green

The pd.Index.get_indexer method can help us convert a label or list of labels into their corresponding positions in a pd.Index:

```
col_idxer = df.columns.get_indexer(["age", "eye_color"])
col_idxer
```

```
array([0, 2])
```

This can subsequently be used as an argument to .iloc, ensuring that you use position-based selection across both the rows and columns:

```
df.iloc[[0, 1], col_idxer]
```

	age	eye_color
0	24	blue
1	42	brown

There's more...

Instead of using pd.Index.get_indexer, you can split this expression up into a few steps, with one of the steps performing index-based selection and the other performing label-based selection. And if you did this, you'd end up getting the exact same result as shown above:

```
df[["age", "eye_color"]].iloc[[0, 1]]
```

	age	eye_color
0	24	blue
1	42	brown

There's a strong argument to be made that this is more expressive than using pd.Index.get_indexer, which developers of all experience levels with pandas would agree with. So why even bother with pd.Index.get_indexer?

While these appear the same on the surface, how pandas computes the result is drastically different. Adding some timing benchmarks to the various methods should highlight this. While the exact numbers will vary on your machine, compare the timing output of the idiomatic approach described in this section:

```
import timeit

def get_indexer_approach():
  col_idxer = df.columns.get_indexer(["age", "eye_color"])
  df.iloc[[0, 1], col_idxer]

timeit.timeit(get_indexer_approach, number=10_000)
```

```
1.8184850879988517
```

to the approach with separate steps to select by label and then by position:

```
two_step_approach = lambda: df[["age", "eye_color"]].iloc[[0, 1]]
timeit.timeit(two_step_approach, number=10_000
```

```
2.027099569000711
```

The pd.Index.get_indexer approach clocks in faster and should scale better to larger datasets. The reason for this is that pandas evaluates its expressions *eagerly* or, more specifically, it will do *what you say, when you say it*. The expression df[["age", "eye_color"]].iloc[[0, 1]] first runs df[["age", "eye_color"]], which creates an intermediate pd.DataFrame, to which the .iloc[[0, 1]] gets applied. By contrast, the expression df.iloc[[0, 1], col_idxer] performs the label-based and position-based selection all in one go, avoiding the creation of any intermediate pd.DataFrame.

The contrasting approach to the *eager execution* approach that pandas takes is often called *lazy execution*. If you've used SQL before, the latter is a good example of that; you typically do not instruct the SQL engine on what steps to take exactly to produce the desired result. Instead, you *declare* what you want your result to look like and leave it up to the SQL database to optimize and execute your query.

Will pandas ever support lazy evaluation and optimization? I would posit yes, as it would help pandas scale to larger datasets and take the onus away from the end user to write optimal queries. However, that capability does not exist today, so it is still important for you as a user of the library to understand if the code you produce will be processed efficiently or inefficiently.

It is also worth considering the context of your data analysis when deciding if it is worth trying to combine position/label-based selection in one step, or if they are fine as separate steps. In our trivial example, the runtime difference of df.iloc[[0, 1], col_idxer] versus df[["age", "eye_color"]]. iloc[[0, 1]] is probably not worth caring about in the grander scheme of things, but if you were dealing with larger datasets and bottlenecked by performance, the former approach could be a lifesaver.

DataFrame.filter

pd.DataFrame.filter is a specialized method that allows you to select from either the rows or columns of a pd.DataFrame.

How to do it

Let's create a pd.DataFrame where we have indices composed of strings in both the rows and columns:

```python
df = pd.DataFrame([
    [24, 180, "blue"],
    [42, 166, "brown"],
    [22, 160, "green"],
], columns=[
    "age",
    "height_cm",
    "eye_color"
], index=["Jack", "Jill", "Jayne"])
df
```

```
       age    height_cm    eye_color
Jack    24    180          blue
Jill    42    166          brown
Jayne   22    160          green
```

By default, pd.DataFrame.filter will select columns matching the label argument(s), similar to pd.DataFrame[]:

```python
df.filter(["age", "eye_color"])
```

```
       age    eye_color
Jack    24    blue
Jill    42    brown
Jayne   22    green
```

However, pd.DataFrame.filter also accepts an axis= argument, which allows you to change the axis being selected from. To select rows instead of columns, pass axis=0:

```python
df.filter(["Jack", "Jill"], axis=0)
```

```
        age    height_cm    eye_color
Jack    24       180          blue
Jill    42       166          brown
```

You are not limited to exact string matches against labels. If you would like to select any label containing a string, use the `like=` parameter. This example will select any column containing an underscore:

```
df.filter(like="_")
```

```
         height_cm    eye_color
Jack     180            blue
Jill     166            brown
Jayne    160            green
```

If simple string containment is not enough, you can also use regular expressions to match index labels with the `regex=` parameter. The following example will select any row labels that start with a `"Ja"` but do not end with `"e"`:

```
df.filter(regex=r"^Ja.*(?<!e)$", axis=0)
```

```
        age    height_cm    eye_color
Jack    24       180          blue
```

Selection by data type

So far in this cookbook, we have *seen* data types, but we have not talked too much in depth about what they are. We still aren't quite there; a deep dive into the type system of pandas is reserved for *Chapter 3, Data Types*. However, for now, you should be aware that the column type provides metadata that `pd.DataFrame.select_dtypes` can use for selection.

How to do it

Let's start with a `pd.DataFrame` that uses integral, floating point, and string columns:

```
df = pd.DataFrame([
    [0, 1.0, "2"],
    [4, 8.0, "16"],
], columns=["int_col", "float_col", "string_col"])
df
```

```
    int_col    float_col    string_col
0    0           1.0          2
1    4           8.0          16
```

Use `pd.DataFrame.select_dtypes` to select only integral columns:

```
df.select_dtypes("int")
```

```
      int_col
0     0
1     4
```

Multiple types can be selected if you pass a list argument:

```
df.select_dtypes(include=["int", "float"])
```

```
      int_col    float_col
0     0          1.0
1     4          8.0
```

The default behavior is to include the data types you pass in as an argument. To exclude them, use the exclude= parameter instead:

```
df.select_dtypes(exclude=["int", "float"])
```

```
      string_col
0     2
1     16
```

Selection/filtering via Boolean arrays

Using Boolean lists/arrays (also referred to as *masks*) is a very common method to select a subset of rows.

How to do it

Let's create a mask of True=/=False values alongside a simple pd.Series:

```
mask = [True, False, True]
ser = pd.Series(range(3))
ser
```

```
0     0
1     1
2     2
dtype: int64
```

Using the mask as an argument to pd.Series[] will return each row where the corresponding mask entry is True:

```
ser[mask]
```

```
0     0
2     2
dtype: int64
```

pd.Series.loc will match the exact same behavior as pd.Series[] in this particular case:

```
ser.loc[mask]
```

```
0    0
2    2
dtype: int64
```

Interestingly, whereas pd.DataFrame[] usually tries to select from the columns when provided a list argument, its behavior with a sequence of Boolean values is different. Using the mask we have already created, df[mask] will actually match along the rows rather than the columns:

```
df = pd.DataFrame(np.arange(6).reshape(3, -1))
df[mask]
```

```
     0   1
0    0   1
2    4   5
```

If you need to mask the columns alongside the rows, pd.DataFrame.loc will accept two mask arguments:

```
col_mask = [True, False]
df.loc[mask, col_mask]
```

```
     0
0    0
2    4
```

There's more...

Commonly, you will manipulate your masks using some combination of the OR, AND, or INVERT operators. To see these in action, let's start with a slightly more complicated pd.DataFrame:

```
df = pd.DataFrame([
    [24, 180, "blue"],
    [42, 166, "brown"],
    [22, 160, "green"],
], columns=["age", "height_cm", "eye_color"], index=["Jack", "Jill", "Jayne"])
df
```

```
       age   height_cm   eye_color
Jack   24    180         blue
Jill   42    166         brown
Jayne  22    160         green
```

If our goal was to filter this only to users with blue or green eyes, we could first identify which users have blue eyes:

```
blue_eyes = df["eye_color"] == "blue"
blue_eyes
```

```
Jack       True
Jill       False
Jayne      False
Name: eye_color, dtype: bool
```

Then, we figure out who has green eyes:

```
green_eyes = df["eye_color"] == "green"
green_eyes
```

```
Jack       False
Jill       False
Jayne      True
Name: eye_color, dtype: bool
```

and combine those together into one Boolean *mask* using the OR operator, |:

```
mask = blue_eyes | green_eyes
mask
```

```
Jack       True
Jill       False
Jayne      True
Name: eye_color, dtype: bool
```

before passing that mask in as an indexer of our pd.DataFrame:

```
df[mask]
```

```
        age    height_cm    eye_color
Jack    24     180          blue
Jayne   22     160          green
```

Instead of using the OR operator, |, you will often commonly use the AND operator, &. For example, let's create a filter for records with an age less than 40:

```
age_lt_40 = df["age"] < 40
age_lt_40
```

```
Jack       True
Jill       False
```

```
Jayne     True
Name: age, dtype: bool
```

And also, a height greater than 170:

```
height_gt_170 = df["height_cm"] > 170
height_gt_170
```

```
Jack      True
Jill      False
Jayne     False
Name: height_cm, dtype: bool
```

These can be ANDed together to only select records that meet both conditions:

```
df[age_lt_40 & height_gt_170]
```

```
      age   height_cm   eye_color
Jack  24    180         blue
```

The INVERT operator is useful to think of as a NOT operator; i.e., in the context of a mask, it will make any `True` value `False` and any `False` value `True`. Continuing with our example above, if we wanted to find records that did not satisfy the condition of having an age under 40 and a height over 170, we could simply invert our mask using ~:

```
df[~(age_lt_40 & height_gt_170)]
```

```
       age   height_cm   eye_color
Jill   42    166         brown
Jayne  22    160         green
```

Selection with a MultiIndex — A single level

A `pd.MultiIndex` is a subclass of a `pd.Index` that supports hierarchical labels. Depending on who you ask, this can be one of the best or one of the worst features of pandas. After reading this cookbook, I hope you consider it one of the best.

Much of the derision toward the `pd.MultiIndex` comes from the fact that the syntax used to select from it can easily become ambiguous, especially when using `pd.DataFrame[]`. The examples below exclusively use the `pd.DataFrame.loc` method and avoid `pd.DataFrame[]` to mitigate confusion.

How to do it

`pd.MultiIndex.from_tuples` can be used to construct a `pd.MultiIndex` from a list of tuples. In the following example, we create a `pd.MultiIndex` with two levels – `first_name` and `last_name`, sequentially. We will pair this alongside a very simple `pd.Series`:

```
index = pd.MultiIndex.from_tuples([
    ("John", "Smith"),
```

```
    ("John", "Doe"),
    ("Jane", "Doe"),
    ("Stephen", "Smith"),
], names=["first_name", "last_name"])
ser = pd.Series(range(4), index=index)
ser
```

```
first_name  last_name
John        Smith        0
            Doe          1
Jane        Doe          2
Stephen     Smith        3
dtype: int64
```

Using pd.Series.loc with a pd.MultiIndex and a scalar argument will match against the first level of the pd.MultiIndex. The output will not include this first level in its result:

```
ser.loc["John"]
```

```
last_name
Smith    0
Doe      1
dtype: int64
```

The behavior that drops the first level of the pd.MultiIndex in the above example is also referred to as *partial slicing*. This concept is similar to the dimensionality reduction we saw with .loc and .iloc in the previous sections, with the exception that instead of reducing *dimensions*, pandas here tries to reduce the number of *levels* in a pd.MultiIndex.

To prevent this implicit level reduction from occurring, we can once again provide a list argument containing a single element:

```
ser.loc[["John"]]
```

```
first_name  last_name
John        Smith        0
            Doe          1
dtype: int64
```

Selection with a MultiIndex — Multiple levels

Things would not be that interesting if you could only select from the first level of a pd.MultiIndex. Fortunately, pd.DataFrame.loc will scale out to more than just the first level through the creative use of tuple arguments.

How to do it

Let's recreate the pd.Series from the previous section:

```
index = pd.MultiIndex.from_tuples([
    ("John", "Smith"),
    ("John", "Doe"),
    ("Jane", "Doe"),
    ("Stephen", "Smith"),
], names=["first_name", "last_name"])
ser = pd.Series(range(4), index=index)
ser
```

```
first_name  last_name
John        Smith        0
            Doe          1
Jane        Doe          2
Stephen     Smith        3
dtype: int64
```

To select all records where the first index level uses the label "Jane" and the second uses "Doe", pass the following tuple:

```
ser.loc[("Jane", "Doe")]
```

```
2
```

To select all records where the first index level uses the label "Jane" and the second uses "Doe", while maintaining the pd.MultiIndex shape, place a single element list in the tuple:

```
ser.loc[(["Jane"], "Doe")]
```

```
first_name  last_name
Jane        Doe          2
dtype: int64
```

To select all records where the first index level uses the label "John" and the second uses the label "Smith", OR the first level is "Jane" and the second is "Doe":

```
ser.loc[[("John", "Smith"), ("Jane", "Doe")]]
```

```
first_name  last_name
John        Smith        0
Jane        Doe          2
dtype: int64
```

To select all records where the second index level is "Doe", use an empty slice as the first tuple element. Note that this drops the second index level and reconstructs the result with a simple pd.Index from the first index level that remains:

```
ser.loc[(slice(None), "Doe")]
```

```
first_name
John    1
Jane    2
dtype: int64
```

To select all records where the second index level is "Doe" while maintaining the pd.MultiIndex shape, pass a single-element list as the second tuple element:

```
ser.loc[(slice(None), ["Doe"])]
```

```
first_name  last_name
John        Doe          1
Jane        Doe          2
dtype: int64
```

At this point, you might be asking yourself the question, what the heck does slice(None) mean? This rather cryptic expression actually creates a slice object without a *start*, *stop*, or *step* value, which is easier to illustrate with a simpler Python list – note that the behavior here:

```
alist = list("abc")
alist[:]
```

```
['a', 'b', 'c']
```

is exactly the same as with slice(None):

```
alist[slice(None)]
```

```
['a', 'b', 'c']
```

When a pd.MultiIndex expects a tuple argument but doesn't get one, this issue is caused by a slice within a tuple, similar to how (:,) is a syntax error in Python. The more explicit (slice(None),) fixes the issue.

There's more...

If you find the slice(None) syntax to be unwieldy, pandas provides a convenient object called the pd.IndexSlice that acts like a tuple but allows you to use the more natural : notation for slicing.

```
ser.loc[(slice(None), ["Doe"])]
```

```
first_name  last_name
John        Doe          1
```

```
Jane          Doe            2
dtype: int64
```

thus can become:

```
ixsl = pd.IndexSlice
ser.loc[ixsl[:, ["Doe"]]]
```

```
first_name    last_name
John          Doe            1
Jane          Doe            2
dtype: int64
```

Selection with a MultiIndex — a DataFrame

A pd.MultiIndex can be used both as a row index and a column index, and selection via pd.DataFrame.loc works with both.

How to do it

Let's create a pd.DataFrame that uses a pd.MultiIndex in both the rows and columns:

```
row_index = pd.MultiIndex.from_tuples([
    ("John", "Smith"),
    ("John", "Doe"),
    ("Jane", "Doe"),
    ("Stephen", "Smith"),
], names=["first_name", "last_name"])
col_index = pd.MultiIndex.from_tuples([
    ("music", "favorite"),
    ("music", "last_seen_live"),
    ("art", "favorite"),
], names=["art_type", "category"])
df = pd.DataFrame([
    ["Swift", "Swift", "Matisse"],
    ["Mozart", "T. Swift", "Van Gogh"],
    ["Beatles", "Wonder", "Warhol"],
    ["Jackson", "Dylan", "Picasso"],
], index=row_index, columns=col_index)
df
```

	art_type		music	art
	category	favorite	last_seen_live	favorite
first_name	last_name			
John	Smith	Swift	Swift	Matisse

		Doe	Mozart	T. Swift	Van Gogh
Jane		Doe	Beatles	Wonder	Warhol
Stephen		Smith	Jackson	Dylan	Picasso

To select all rows where the second level is "Smith" and all columns where the second level is "favorite", you will need to pass two tuples where the second element in each is the desired label:

```python
row_idxer = (slice(None), "Smith")
col_idxer = (slice(None), "favorite")
df.loc[row_idxer, col_idxer]
```

		art_type	music	art
		category	favorite	favorite
first_name	last_name			
John	Smith		Swift	Matisse
Stephen	Smith		Jackson	Picasso

pd.DataFrame.loc always requires two arguments – the first to specify how the rows should be indexed and the second to specify how the columns should be indexed. When you have a pd.DataFrame with a pd.MultiIndex in both the rows and the columns, you may find it stylistically easier to create separate variables for the indexers. The above code could have also been written as:

```python
df.loc[(slice(None), "Smith"), (slice(None), "favorite")]
```

		art_type	music	art
		category	favorite	favorite
first_name	last_name			
John	Smith		Swift	Matisse
Stephen	Smith		Jackson	Picasso

Although you could argue that this is more difficult to interpret. As the old saying goes, beauty is in the eye of the beholder.

Item assignment with .loc and .iloc

The pandas library is optimized for reading, exploring, and evaluating data. Operations that try to *mutate* or change data are far less efficient.

However, when you must mutate your data, you can use .loc and .iloc to do it.

How to do it

Let's start with a very small pd.Series:

```python
ser = pd.Series(range(3), index=list("abc"))
```

pd.Series.loc is useful when you want to assign a value by matching against the label of an index. For example, if we wanted to store the value 42 where our row index contained a value of "b", we would write:

```
ser.loc["b"] = 42
ser
```

```
a     0
b    42
c     2
dtype: int64
```

pd.Series.iloc is used when you want to assign a value positionally. To assign the value -42 to the second element in our pd.Series, we would write:

```
ser.iloc[2] = -42
ser
```

```
a     0
b    42
c   -42
dtype: int64
```

There's more...

The cost of mutating data through pandas can depend largely on two factors:

- The type of array backing a pandas pd.Series (*Chapter 3*, *Data Types*, will cover data types in more detail)
- How many objects reference a pd.Series

A deep dive into those factors is far beyond the scope of this book. For the first point above, my general guidance is that the *simpler* an array type is, the better your odds are of being able to mutate it without the array contents having to be copied, which for larger datasets may be prohibitively expensive.

For the second bullet, a lot of **Copy on Write (CoW)** work was involved in the pandas 2.x series. CoW is the default behavior in pandas 3.0, and it tries to make the behavior of what does and does not get copied when mutating data more predictable. For advanced users, I highly encourage giving the pandas CoW documentation a read.

DataFrame column assignment

While assigning to *data* can be a relatively expensive operation in pandas, assigning columns to a pd.DataFrame is a common operation.

How to do it

Let's create a very simple pd.DataFrame:

```
df = pd.DataFrame({"col1": [1, 2, 3]})
df
```

```
     col1
0    1
1    2
2    3
```

New columns can be assigned using the pd.DataFrame[] operator. The simplest type of assignment can take a scalar value and *broadcast* it to every row of the pd.DataFrame:

```
df["new_column1"] = 42
df
```

```
     col1    new_column1
0    1       42
1    2       42
2    3       42
```

You can also assign a pd.Series or sequence as long as the number of elements matches the number of rows in the pd.DataFrame:

```
df["new_column2"] = list("abc")
df
```

```
     col1    new_column1    new_column2
0    1       42             a
1    2       42             b
2    3       42             c
```

```
df["new_column3"] = pd.Series(["dog", "cat", "human"])
df
```

```
     col1    new_column1    new_column2    new_column3
0    1       42             a              dog
1    2       42             b              cat
2    3       42             c              human
```

If the new sequence does not match the number of rows in the existing pd.DataFrame, the assignment will fail:

```
df["should_fail"] = ["too few", "rows"]
```

```
ValueError: Length of values (2) does not match length of index (3)
```

Assignment can also be done to a pd.DataFrame with a pd.MultiIndex in the columns. Let's take a look at such a pd.DataFrame:

```
row_index = pd.MultiIndex.from_tuples([
    ("John", "Smith"),
    ("John", "Doe"),
```

```
        ("Jane", "Doe"),
        ("Stephen", "Smith"),
    ], names=["first_name", "last_name"])
    col_index = pd.MultiIndex.from_tuples([
        ("music", "favorite"),
        ("music", "last_seen_live"),
        ("art", "favorite"),
    ], names=["art_type", "category"])
    df = pd.DataFrame([
        ["Swift", "Swift", "Matisse"],
        ["Mozart", "T. Swift", "Van Gogh"],
        ["Beatles", "Wonder", "Warhol"],
        ["Jackson", "Dylan", "Picasso"],
    ], index=row_index, columns=col_index)
    df
```

	art_type	music	art	
	category	favorite	last_seen_live	favorite
first_name	last_name			
John	Smith	Swift	Swift	Matisse
	Doe	Mozart	T. Swift	Van Gogh
Jane	Doe	Beatles	Wonder	Warhol
Stephen	Smith	Jackson	Dylan	Picasso

To assign a new column under the "art" hierarchy for the number of museums seen, pass a tuple argument to pd.DataFrame.loc:

```
    df.loc[:, ("art", "museuems_seen")] = [1, 2, 4, 8]
    df
```

	art_type	music	art		
	category	favorite	last_seen_live	favorite	museuems_seen
first_name	last_name				
John	Smith	Swift	Swift	Matisse	1
	Doe	Mozart	T. Swift	Van Gogh	2
Jane	Doe	Beatles	Wonder	Warhol	4
Stephen	Smith	Jackson	Dylan	Picasso	8

Assignment with a pd.DataFrame follows the same patterns we saw when selecting values with pd.DataFrame[] and pd.DataFrame.loc[]. The main difference is that during selection, you would use pd.DataFrame[] and pd.DataFrame.loc[] on the right-hand side of an expression, whereas with assignment, they appear on the left-hand side.

There's more...

The `pd.DataFrame.assign` method can be used to allow *method chaining* during assignment. Let's start with a simple `pd.DataFrame` to illustrate the utility:

```
df = pd.DataFrame([[0, 1], [2, 4]], columns=list("ab"))
df
```

```
     a   b
0    0   1
1    2   4
```

Method chaining refers to the ability of pandas to apply many algorithms in succession to a pandas data structure (algorithms and how to apply them will be covered in more detail in *Chapter 5, Algorithms and How to Apply Them*). So, to take our `pd.DataFrame`, double it, and add 42 to each element, we could do something like:

```
(
    df
    .mul(2)
    .add(42)
)
```

```
     a    b
0    42   44
1    46   50
```

But what happens if we want to add a new column as part of this chain of events? Unfortunately, with the standard assignment operators, you would have to break that chain of events and usually assign a new variable:

```
df2 = (
    df
    .mul(2)
    .add(42)
)
df2["assigned_c"] = df2["b"] - 3
df2
```

```
     a    b    assigned_c
0    42   44   41
1    46   50   47
```

But with `pd.DataFrame.assign`, you can continue chaining along. Simply pass the desired column label as a keyword to `pd.DataFrame.assign`, whose argument is the values you would like to see in the new `pd.DataFrame`:

```
(
    df
    .mul(2)
    .add(42)
    .assign(chained_c=lambda df: df["b"] - 3)
)
```

	a	b	chained_c
0	42	44	41
1	46	50	47

In this case, you are limited to using labels that meet Python's syntax requirements for parameter names, and this, unfortunately, does not work with a pd.MultiIndex. Some users think method chaining makes debugging harder, while others argue that method chaining like this makes code easier to read. Ultimately, there is no right or wrong answer, and the best advice I can give you for now is to use the form you feel most comfortable with.

3

Data Types

The data type of a `pd.Series` allows you to dictate what kind of elements may or may not be stored. Data types are important for ensuring data quality, as well as enabling high-performance algorithms in your code. If you have a data background working with databases, you more than likely are already familiar with data types and their benefits; you will find types like TEXT, INTEGER, and DOUBLE PRECISION in pandas just like you do in a database, albeit under different names.

Unlike a database, however, pandas offers multiple implementations of how a TEXT, INTEGER, and DOUBLE PRECISION type can work. Unfortunately, this means, as an end user, that you should at least have some understanding of how the different data types are implemented to make the best choice for your application.

A quick history lesson on types in pandas can help explain this usability quirk. Originally, pandas was built on top of the NumPy type system. This worked for quite a while but had major shortcomings. For starters, the NumPy types pandas built on top of did not support missing values, so pandas created a Frankenstein's monster of methods to support those. NumPy, being focused on *numerical* computations, also did not offer a first-class string data type, leading to very poor string handling in pandas.

Work to move past the NumPy type system started with pandas version 0.23, which introduced new data types built directly into pandas that were still implemented using NumPy but could actually handle missing values. In version 1.0, pandas implemented its own string data type. At the time, these were called `numpy_nullable` data types, but over time, they have become referred to as pandas extension types.

While all of this was going on, Wes McKinney, the original creator of pandas, was working on the Apache Arrow project. Fully explaining the Arrow project is beyond the scope of this book, but one of the major things it helps with is to define a set of standardized data types that can be used from different tools and programming languages. Those data types also draw inspiration from databases; if using a database has already been a part of your analytics journey, then the Arrow types will likely be very familiar to you. Starting with version 2.0, pandas allows you to use Arrow for your data types.

Despite support for pandas extension and Arrow data types, the default types from pandas were never changed, and in most cases still use NumPy. In the author's opinion, this is very unfortunate; this chapter will introduce a rather opinionated take on how to best manage the type landscape, with the general guidance of the following:

- Use pandas extension types, when available
- Use Arrow data types, when pandas extension types do not suffice
- Use NumPy-backed data types

This guidance may be controversial and can be scrutinized in extreme examples, but, for someone just starting with pandas, I believe this prioritization gives users the best balance of usability and performance, without requiring a deep understanding of how pandas works behind the scenes.

The general layout of this chapter will introduce the pandas extension system for general use, before diving into the Arrow-type system for more advanced use cases. As we walk through these types, we will also highlight any special behavior that can be unlocked using *accessors*. Finally, we will talk about the historical NumPy-backed data types and take a deep dive into some of their fatal flaws, which I hope will convince you as to why you should limit your use of these types.

We are going to cover the following recipes in this chapter:

- Integral types
- Floating point types
- Boolean types
- String types
- Missing value handling
- Categorical types
- Temporal types – datetime
- Temporal types – timedelta
- Temporal PyArrow types
- PyArrow List types
- PyArrow decimal types
- NumPy type system, the object type, and pitfalls

Integral types

Integral types are the most basic type category. Much like the int type in Python or the INTEGER data type in a database, these can only represent whole numbers. Despite this limitation, integers are useful in a wide variety of applications, including but not limited to arithmetic, indexing, counting, and enumeration.

Integral types are heavily optimized for performance, tracing all the way from pandas down to the hardware on your computer. The integral types offered by pandas are significantly faster than the int type offered by the Python standard library, and proper usage of integral types is often a key enabler to high-performance, scalable reporting.

How to do it

Any valid sequence of integers can be passed as an argument to the pd.Series constructor. Paired with the dtype=pd.Int64Dtype() argument you will end up with a 64-bit integer data type:

```
pd.Series(range(3), dtype=pd.Int64Dtype())
```

```
0    0
1    1
2    2
dtype: Int64
```

When storage and compute resources are not a concern, users often opt for 64-bit integers, but we could have also picked a smaller data type in our example:

```
pd.Series(range(3), dtype=pd.Int8Dtype())
```

```
0    0
1    1
2    2
dtype: Int8
```

With respect to missing values, pandas uses the pd.NA sentinel as its indicator, much like a database uses NULL:

```
pd.Series([1, pd.NA, 2], dtype=pd.Int64Dtype())
```

```
0       1
1    <NA>
2       2
dtype: Int64
```

As a convenience, the pd.Series constructor will convert Python None values into pd.NA for you:

```
pd.Series([1, None, 2], dtype=pd.Int64Dtype())
```

```
0       1
1    <NA>
2       2
dtype: Int64
```

There's more...

For users new to scientific computing, it is important to know that unlike Python's int, which has no theoretical size limit, integers in pandas have lower and upper bounds. These limits are determined by the *width* and *signedness* of the integer.

In most computing environments, users have integer widths of 8, 16, 32, and 64. Signedness can be either *signed* (i.e., the number can be positive or negative) or *unsigned* (i.e., the number must not be negative). Limits for each integral type are summarized in the following table:

Type	Lower Bound	Upper Bound
8-bit width, signed	-128	127
8-bit width, unsigned	0	255
16-bit width, signed	-32769	32767
16-bit width, unsigned	0	65535
32-bit width, signed	-2147483648	2147483647
32-bit width, unsigned	0	4294967295
64-bit width, signed	-(2**63)	2**63-1
64-bit width, unsigned	0	2**64-1

Table 3.1: Integral limits per signedness and width

The trade-off in these types is capacity versus memory usage – a 64-bit integral type requires 8x as much memory as an 8-bit integral type. Whether or not this is an issue depends entirely on the size of your dataset and the system on which you perform your analysis.

Within the pandas extension type system, the `dtype=` argument for each of these follows the `pd.IntXXDtype()` form for signed integers and `pd.UIntXXDtype()` for unsigned integers, where `XX` refers to the bit width:

```
pd.Series(range(555, 558), dtype=pd.Int16Dtype())
```

```
0    555
1    556
2    557
dtype: Int16
```

```
pd.Series(range(3), dtype=pd.UInt8Dtype())
```

```
0    0
1    1
2    2
dtype: UInt8
```

Floating point types

Floating point types allow you to represent real numbers, not just integers. This allows you to work with a continuous and *theoretically* infinite set of values within your computations. It may come as no surprise that floating point calculations show up in almost every scientific computation, macro-financial analysis, machine learning algorithm, and so on.

The emphasis on the word *theoretically*, however, is intentional and very important to understand. Floating point types still have boundaries, with real limitations being imposed by your computer hardware. In essence, the notion of being able to represent any number is an illusion. Floating point types are liable to lose precision and introduce rounding errors, especially as you work with more extreme values. As such, floating point types are not suitable when you need absolute precision (for that, you will want to reference the PyArrow decimal types introduced later in this chapter).

Despite those limitations, it is rare that you actually would need absolute precision, so floating point types are the most commonly used data type to represent fractional numbers in general.

How to do it

To construct floating point data, use `dtype=pd.Float64Dtype()`:

```
pd.Series([3.14, .333333333, -123.456], dtype=pd.Float64Dtype())
```

```
0         3.14
1     0.333333
2     -123.456
dtype: Float64
```

Much like we saw with the integral types, the missing value indicator is `pd.NA`. The Python object `None` will be implicitly converted to this as a convenience:

```
pd.Series([3.14, None, pd.NA], dtype=pd.Float64Dtype())
```

```
0     3.14
1     <NA>
2     <NA>
dtype: Float64
```

There's more...

By nature of their design, floating point values are *inexact*, and arithmetic with floating point values is slower than with integers. A deep dive into floating point arithmetic is beyond the scope of this book, but those interested can find much more information in the Python documentation.

Python has a built-in `float` type that is somewhat of a misnomer because it is actually an IEEE 754 `double`. That standard and other languages like C/C++ have distinct `float` and `double` types, with the former occupying 32 bits and the latter occupying 64 bits. To disambiguate these widths but stay consistent with Python terminology, pandas offers `pd.Float64Dtype()` (which some may consider a `double`) and `pd.Float32Dtype()` (which some may consider a `float`).

Generally, unless your system is constrained on resources, users are recommended to use 64-bit floating point types. The odds of losing precision with 32-bit floating point types are much higher than with their respective 64-bit counterparts. In fact, 32-bit floats only offer between 6 and 9 decimal digits of precision, so the following expression will likely return `True` for equality comparison, even though we as humans can very clearly see the numbers are not the same:

```
ser1 = pd.Series([1_000_000.123], dtype=pd.Float32Dtype())
ser2 = pd.Series([1_000_000.124], dtype=pd.Float32Dtype())
ser1.eq(ser2)
```

```
0    True
dtype: boolean
```

With a 64-bit floating point, you would at least get between 15 and 17 decimal digits of precision, so the values at which rounding errors occur are much more extreme.

Boolean types

A Boolean type represents a value that is either `True` or `False`. Boolean data types are useful to simply answer questions with a yes/no style of response and are also widely used in machine learning algorithms to convert categorical values into 1s and 0s (for `True` and `False`, respectively) that a computer can more easily digest (see also the *One-hot encoding with pd.get_dummies* recipe in *Chapter 5, Algorithms and How to Apply Them*).

How to do it

For Boolean, the appropriate `dtype=` argument is `pd.BooleanDtype`:

```
pd.Series([True, False, True], dtype=pd.BooleanDtype())
```

```
0     True
1    False
2     True
dtype: boolean
```

The pandas library will take care of implicitly converting values to their Boolean representation for you. Often, 0 and 1 are used in place of `False` and `True`, respectively:

```
pd.Series([1, 0, 1], dtype=pd.BooleanDtype())
```

```
0     True
1    False
2     True
dtype: boolean
```

Once again, pd.NA is the canonical missing indicator, although pandas will implicitly convert None to a missing value:

```
pd.Series([1, pd.NA, None], dtype=pd.BooleanDtype())
```

```
0       True
1       <NA>
2       <NA>
dtype: boolean
```

String types

The string data type is the appropriate choice for any data that represents text. Unless you are working in a purely scientific domain, chances are that strings will be prevalent throughout the data that you use.

In this recipe, we will highlight some of the additional features pandas provides when working with string data, most notably through the pd.Series.str accessor. This accessor helps to change cases, extract substrings, match patterns, and more.

As a technical note, before we jump into the recipe, strings starting in pandas 3.0 will be significantly overhauled behind the scenes, enabling an implementation that is more type-correct, much faster, and requires far less memory than what was available in the pandas 2.x series. To make this possible in 3.0 and beyond, users are highly encouraged to install PyArrow alongside their pandas installation. For users looking for an authoritative reference on the why and how of strings in pandas 3.0, you may reference the PDEP-14 dedicated string data type.

How to do it

String data should be constructed with dtype=pd.StringDtype():

```
pd.Series(["foo", "bar", "baz"], dtype=pd.StringDtype())
```

```
0       foo
1       bar
2       baz
dtype: string
```

You have probably picked up by now that pd.NA is the missing indicator to use, but pandas will convert None implicitly for you:

```
pd.Series(["foo", pd.NA, None], dtype=pd.StringDtype())
```

```
0       foo
1       <NA>
2       <NA>
dtype: string
```

When working with a pd.Series containing string data, pandas will create what it refers to as the string *accessor* to help you unlock new methods that are tailored to strings. The string accessor is used via pd.Series.str, and helps you do things like report back the length of each string via pd.Series. str.len:

```
ser = pd.Series(["xx", "YyY", "zZzZ"], dtype=pd.StringDtype())
ser.str.len()
```

```
0    2
1    3
2    4
dtype: Int64
```

It may also be used to force everything to a particular case, like uppercase:

```
ser.str.upper()
```

```
0      XX
1     YYY
2    ZZZZ
dtype: string
```

It may also be used to force everything to lowercase:

```
ser.str.lower()
```

```
0      xx
1     yyy
2    zzzz
dtype: string
```

And even "title case" (i.e., the first letter only is capitalized, with everything else lower):

```
ser.str.title()
```

```
0      Xx
1     Yyy
2    Zzzz
dtype: string
```

pd.Series.str.contains can be used to check for simple string containment:

```
ser = pd.Series(["foo", "bar", "baz"], dtype=pd.StringDtype())
ser.str.contains("o")
```

```
0       True
1       False
2       False
dtype: boolean
```

But it also has the flexibility to test for regular expressions with `regex=True`, akin to how `re.search` works in the standard library. The `case=False` argument will also turn the matching into a case-insensitive comparison:

```
ser.str.contains(r"^ba[rz]$", case=False, regex=True)
```

```
0       False
1       True
2       True
dtype: boolean
```

Missing value handling

Before we continue with more data types, we must step back and talk about how pandas handles missing values. So far, things have been simple (we have only seen `pd.NA`), but as we explore more types we will see that the way pandas handles missing values is inconsistent, owing mostly to the history of how the library was developed. While it would be great to wave a magic wand and make any inconsistencies go away, in reality, they have existed and will continue to exist in production code bases for years to come. Having a high-level understanding of that evolution will help you write better pandas code, and hopefully convert the unaware to using the idioms we preach in this book.

How to do it

The pandas library was originally built on top of NumPy, whose default data types do not support missing values. As such, pandas had to build its own missing value handling solution from scratch, and, for better or worse, decided that using the `np.nan` sentinel, which represents "not a number," was useful enough to build off of.

`np.nan` itself is an implementation of the IEEE 754 standard's "not a number" sentinel, a specification that only really had to do with floating point arithmetic. There is no such thing as "not a number" for integral data, which is why pandas implicitly converts a `pd.Series` like this:

```
ser = pd.Series(range(3))
ser
```

```
0       0
1       1
2       2
dtype: int64
```

To a floating point data type after assigning a missing value:

```
ser.iloc[1] = None
ser
```

```
0    0.0
1    NaN
2    2.0
dtype: float64
```

As we discussed back in the *Floating point types* recipe, floating point values are slower than their integral counterparts. While integers can be expressed with 8- and 16-bit widths, floating point types require 32 bits at a minimum. Even if you are using 32-bit width integers, using a 32-bit floating point value may not be viable without loss of precision, and with 64-bit integers, conversion simply may just have to lose precision. Generally, with a conversion from integral to floating point types, you have to sacrifice some combination of performance, memory usage, and/or precision, so such conversions are not ideal.

Of course, pandas offers more than just integral and floating point types, so other types had to have custom missing value solutions attached to them. The default Boolean type gets converted to an `object` type, whose pitfalls will be explored in a recipe toward the end of this chapter. For datetime types, which we will discuss soon, pandas had to create a different `pd.NaT` sentinel altogether, as `np.nan` was technically not a feasible value to use for that data type. In essence, each data type in pandas could have its own indicator and implicit casting rules, which are hard to explain for beginners and seasoned pandas developers alike.

The pandas library tried to solve these issues with the introduction of the *pandas extension types* back in the 0.24 release, and as we have seen with the recipes introduced so far, they do a good job of using just `pd.NA` without any implicit type conversion when missing values appear. However, the *pandas extension types* were introduced as opt-in types instead of being the default, so the custom solutions pandas developed to deal with missing values are still prevalent in code. Without having ever rectified these inconsistencies, it is unfortunately up to the user to understand the data types they choose and how different data types handle missing values.

Despite the inconsistencies, pandas fortunately offers a `pd.isna` function that can tell you whether an element in your array is missing or not. This works with the default data types:

```
pd.isna(pd.Series([1, np.nan, 2]))
```

```
0    False
1     True
2    False
dtype: bool
```

It works just as well as it works with the *pandas extension types*:

```
pd.isna(pd.Series([1, pd.NA, 2], dtype=pd.Int64Dtype()))
```

```
0    False
1     True
2    False
dtype: bool
```

There's more...

Users should be aware that comparisons with np.nan and pd.NA do not behave in the same manner. For instance, np.nan == np.nan returns False, whereas pd.NA == pd.NA returns pd.NA. The former comparison is dictated by the terms of IEEE 757, whereas the pd.NA sentinel follows Kleene logic.

The way pd.NA works allows for much more expressive masking/selection in pandas. For instance, if you wanted to create a Boolean mask that also had missing values and use that to select values, pd.BooleanDtype allows you to do so, and naturally will only select records where the mask is True:

```
ser = pd.Series(range(3), dtype=pd.Int64Dtype())
mask = pd.Series([True, pd.NA, False], dtype=pd.BooleanDtype())
ser[mask]
```

```
0    0
dtype: Int64
```

The equivalent operation without the Boolean extension type will raise an error:

```
mask = pd.Series([True, None, False])
ser[mask]
```

```
ValueError: Cannot mask with non-boolean array containing NA / NaN values
```

So, in code that does not use pd.BooleanDtype, you will likely see a lot of method calls that replace "missing" values with False, and use pd.Series.astype to try and cast back to a Boolean data type after the fill:

```
mask = pd.Series([True, None, False])
mask = mask.fillna(False).astype(bool)
ser[mask]
```

```
/tmp/ipykernel_45649/2987852505.py:2: FutureWarning: Downcasting object dtype
arrays on .fillna, .ffill, .bfill is deprecated and will change in a future
version. Call result.infer_objects(copy=False) instead. To opt-in to the future
behavior, set `pd.set_option('future.no_silent_downcasting', True)`
  mask = mask.fillna(False).astype(bool)
0    0
dtype: Int64
```

This is needlessly complex and inefficient. Using pd.BooleanDtype expresses the intent of your operations much more succinctly, letting you worry less about the nuances of pandas.

Categorical types

The main point of the categorical data type is to define an acceptable set of domain values that your pd.Series can contain. The *CSV - strategies for reading large files* recipe in *Chapter 4, The pandas I/O System*, will show you an example where this can result in significant memory savings, but generally, the use case here is to have pandas convert string values like foo, bar, and baz into codes 0, 1, and 2, respectively, which can be much more efficiently stored.

How to do it

So far, we have always opted for pd.XXDtype() as the dtype= argument, which still *could* work in the case of categorical data types, but unfortunately does not handle missing values consistently (see *There's more...* for a deeper dive into this). Instead, we have to opt for one of two alternative approaches to creating a pd.CategoricalDtype with the pd.NA missing value indicator.

With either approach, you will want to start with a pd.Series of data using pd.StringDtype:

```
values = ["foo", "bar", "baz"]
values_ser = pd.Series(values, dtype=pd.StringDtype())
```

From there, you may use pd.DataFrame.astype to cast this to categorical:

```
ser = values_ser.astype(pd.CategoricalDtype())
ser
```

```
0    foo
1    bar
2    baz
dtype: category
Categories (3, string): [bar, baz, foo]
```

Or, if you need more control over the behavior of the categorical type, you may construct a pd.CategoricalDtype from your pd.Series of values and subsequently use that as the dtype= argument:

```
cat = pd.CategoricalDtype(values_ser)
ser = pd.Series(values, dtype=cat)
ser
```

```
0    foo
1    bar
2    baz
dtype: category
Categories (3, string): [foo, bar, baz]
```

Both approaches get you to the same place, although the second approach trades some verbosity in constructing the pd.CategoricalDtype for finer-grained control over its behavior, as you will see throughout the remainder of this recipe.

Regardless of the approach you take, you should note that the values used at the time you construct your categorical pd.Series define the set of acceptable domain values that can be used. Given that we created our categorical type with values of ["foo", "bar", "baz"], subsequent assignment using any of these values is not a problem:

```
ser.iloc[2] = "foo"
ser
```

```
0      foo
1      bar
2      foo
dtype: category
Categories (3, string): [foo, bar, baz]
```

However, assigning a value outside of that domain will raise an error:

```
ser.iloc[2] = "qux"
```

```
TypeError: Cannot setitem on a Categorical with a new category (qux), set the
categories first
```

When explicitly constructing a pd.CategoricalDtype, you can assign a non-lexicographical order to your values via the ordered= argument. This is invaluable when working with *ordinal* data whose values are not naturally sorted the way you want by a computer algorithm.

As a practical example, let's consider the use case of clothing sizes. Naturally, small clothing is smaller than medium clothing, which is smaller than large clothing, and so on. By constructing pd.CategoricalDtype with the desired sizes in order and using ordered=True, pandas makes it very natural to compare sizes:

```
shirt_sizes = pd.Series(["S", "M", "L", "XL"], dtype=pd.StringDtype())
cat = pd.CategoricalDtype(shirt_sizes, ordered=True)
ser = pd.Series(["XL", "L", "S", "L", "S", "M"], dtype=cat)
ser < "L"
```

```
0      False
1      False
2       True
3      False
4       True
5       True
dtype: bool
```

So, how does pandas make this so easy and efficient? The pandas library exposes a categorical accessor pd.Series.cat, which allows you to understand this more deeply. To dive further into this, let's first create a pd.Series of categorical data where a given category is used more than once:

```
accepted_values = pd.Series(["foo", "bar"], dtype=pd.StringDtype())
```

```
cat = pd.CategoricalDtype(accepted_values)
ser = pd.Series(["foo", "bar", "foo"], dtype=cat)
ser
```

```
0    foo
1    bar
2    foo
dtype: category
Categories (2, string): [foo, bar]
```

If you inspect pd.Series.cat.codes, you will see a like-sized pd.Series, but the value foo is replaced with the number 0, and the value bar is replaced with the value 1:

```
ser.cat.codes
```

```
0    0
1    1
2    0
dtype: int8
```

Separately, pd.Series.cat.categories will contain the values of each category, in order:

```
ser.cat.categories
```

```
Index(['foo', 'bar'], dtype='string')
```

Sparing some details around the internals, you can think of pandas as creating a dictionary of the form {0: "foo", 1: "bar"}. While it internally stores a pd.Series with values of [0, 1, 0], when it comes time to display or access the values in any way, those values are used like keys in a dictionary to access the true value the end user would like to use. For this reason, you will often see the categorical data type described as a dictionary type (Apache Arrow, for one, uses the term dictionary).

So, why bother? The process of *encoding* the labels into very small integer lookup values can have a significant impact on memory usage. Note the difference in memory usage between a normal string type:

```
pd.Series(["foo", "bar", "baz"] * 100, dtype=pd.StringDtype()).memory_usage()
```

```
2528
```

As compared to the equivalent categorical type, as follows:

```
pd.Series(["foo", "bar", "baz"] * 100, dtype=cat).memory_usage()
```

```
552
```

Your numbers may or may not exactly match the output of .memory_usage(), but you should at the very least see a rather drastic reduction when using the categorical data type.

There's more...

If using dtype=pd.CategoricalDtype() works directly, why would users not want to use that? Unfortunately, there is a rather large gap in the pandas API that prevents missing values from propagating with categorical types, which can unexpectedly introduce the np.nan missing value indicator we cautioned against using in the *Missing value handling* recipe. This can lead to very surprising behavior, even if you think you are properly using the pd.NA sentinel:

```
pd.Series(["foo", "bar", pd.NA], dtype=pd.CategoricalDtype())
```

```
0      foo
1      bar
2      NaN
dtype: category
Categories (2, object): ['bar', 'foo']
```

Notice in the preceding example that we tried to supply pd.NA but *still* got an np.nan in return? The explicit construction of a pd.CategoricalDtype from a pd.Series with dtype=pd.StringDtype() helps us avoid this very surprising behavior:

```
values = pd.Series(["foo", "bar"], dtype=pd.StringDtype())
cat = pd.CategoricalDtype(values)
pd.Series(["foo", "bar", pd.NA], dtype=cat)
```

```
0      foo
1      bar
2      <NA>
dtype: category
Categories (2, string): [foo, bar]
```

If you find this behavior confusing or troublesome, trust that you are not alone. The light at the end of the tunnel may be PDEP-16, which aims to make pd.NA exclusively work as the missing value indicator. This would mean that you could safely use the pd.CategoricalDtype() constructor directly and follow all the same patterns you saw up until this point.

Unfortunately, this book was released around the time of the pandas 3.0 release and before PDEP-16 had been officially accepted, so it is hard to see into the future and advise when these inconsistencies in the API will go away. If you are reading this book a few years after publication, be sure to check back on the status of PDEP-16, as it may change the proper way to construct categorical data (alongside other data types).

Temporal types — datetime

The term *temporal* generally encompasses data types that concern themselves with dates and times, both in absolute terms as well as when measuring the duration between two different points in time. Temporal types are a key enabler for time-series-based analyses, which can be invaluable for trend detection and forecasting models. In fact, pandas was initially written at a capital management firm before being open sourced. Much of the time-series handling that was built into pandas has been influenced by real-world reporting needs from financial and economic industries.

While the *Categorical types* section started to show some inconsistencies in the pandas type system API, temporal types take things a bit further. It would be reasonable to expect pd.DatetimeDtype() to exist as a constructor, but that is unfortunately not the case, at least as of writing. Additionally, and as mentioned in the *Missing value handling* recipe, temporal types, which were implemented before the pandas type extension system, use a different missing value indicator of pd.NaT (i.e., "not a time").

Despite these issues, pandas offers a mind-boggling amount of advanced functionality for dealing with temporal data. *Chapter 9, Temporal Data Types and Algorithms*, will dive further into the applications of these data types; for now, we will just give a quick overview.

How to do it

Unlike many database systems that offer separate DATE and DATETIME or TIMESTAMP data types, pandas just has a single "datetime" type, which can be constructed via the dtype= argument of the "datetime64[<unit>]" form.

Through much of the history of pandas, ns was the only accepted value for <unit>, so, let's start with that for now (but check *There's more…* for a more detailed explanation of the different values):

```
ser = pd.Series([
    "2024-01-01 00:00:00",
    "2024-01-02 00:00:01",
    "2024-01-03 00:00:02"
], dtype="datetime64[ns]")
ser
```

```
0    2024-01-01 00:00:00
1    2024-01-02 00:00:01
2    2024-01-03 00:00:02
dtype: datetime64[ns]
```

You can also construct a pd.Series of this data type using string arguments without time components:

```
ser = pd.Series([
    "2024-01-01",
    "2024-01-02",
    "2024-01-03"
```

```
], dtype="datetime64[ns]")
ser
```

```
0    2024-01-01
1    2024-01-02
2    2024-01-03
dtype: datetime64[ns]
```

The output of the preceding construction is slightly misleading; although the timestamps are not displayed, pandas still internally stores these values as datetimes, not dates. This might be problematic because there is no way to prevent subsequent timestamps from being stored in that pd.Series:

```
ser.iloc[1] = "2024-01-04 00:00:42"
ser
```

```
0    2024-01-01 00:00:00
1    2024-01-04 00:00:42
2    2024-01-03 00:00:00
dtype: datetime64[ns]
```

If preserving dates is important, be sure to read the *Temporal PyArrow types* recipe later in this chapter.

Much like we saw back with string types, a pd.Series containing datetime data gets an *accessor*, which unlocks features to fluidly deal with dates and times. In this case, the accessor is pd.Series.dt.

We can use this accessor to determine the year of each element in our pd.Series:

```
ser.dt.year
```

```
0    2024
1    2024
2    2024
dtype: int32
```

pd.Series.dt.month will yield the month:

```
ser.dt.month
```

```
0    1
1    1
2    1
dtype: int32
```

pd.Series.dt.day extracts the day of the month that the date falls on:

```
ser.dt.day
```

```
0    1
1    4
```

```
2    3
dtype: int32
```

There is also a `pd.Series.dt.day_of_week` function, which will tell you the day of the week a date falls on. Monday starts at `0`, going up to `6`, meaning Sunday:

```
ser.dt.day_of_week
```

```
0    0
1    3
2    2
dtype: int32
```

If you've worked with timestamps before (especially in global organizations), another thing you may question is what time these values represent. 2024-01-03 00:00:00 in New York City does not happen simultaneously with 2024-01-03 00:00:00 in London, nor in Shanghai. So, how can we get a *true* representation of time?

The timestamps we have seen before are considered *timezone-naive*, (i.e., they do not clearly represent a single point in time anywhere on Earth). By contrast, you can make your timestamps *timezone-aware* by specifying a timezone as part of the `dtype=` argument.

Strangely enough, pandas does have a `pd.DatetimeTZDtype()`, so we can use that along with a `tz=` argument to specify the time zone in which our events are assumed to occur. For example, to make your timestamps represent UTC, you would do the following:

```
pd.Series([
    "2024-01-01 00:00:01",
    "2024-01-02 00:00:01",
    "2024-01-03 00:00:01"
], dtype=pd.DatetimeTZDtype(tz="UTC"))
```

```
0    2024-01-01 00:00:01+00:00
1    2024-01-02 00:00:01+00:00
2    2024-01-03 00:00:01+00:00
dtype: datetime64[ns, UTC]
```

The string UTC represents an **Internet Assigned Numbers Authority (IANA)** timezone identifier. You can use any of those identifiers as the `tz=` argument, like `America/New_York`:

```
pd.Series([
    "2024-01-01 00:00:01",
    "2024-01-02 00:00:01",
    "2024-01-03 00:00:01"
], dtype=pd.DatetimeTZDtype(tz="America/New_York"))
```

```
0    2024-01-01 00:00:01-05:00
1    2024-01-02 00:00:01-05:00
2    2024-01-03 00:00:01-05:00
dtype: datetime64[ns, America/New_York]
```

In case you did not want to use a timezone identifier, you could alternatively specify a UTC offset:

```
pd.Series([
    "2024-01-01 00:00:01",
    "2024-01-02 00:00:01",
    "2024-01-03 00:00:01"
], dtype=pd.DatetimeTZDtype(tz="-05:00"))
```

```
0    2024-01-01 00:00:01-05:00
1    2024-01-02 00:00:01-05:00
2    2024-01-03 00:00:01-05:00
dtype: datetime64[ns, UTC-05:00]
```

The pd.Series.dt accessor we introduced in this recipe also has some nice features for working with timezones. For instance, if you are working with data that technically has no timezone associated with it, but you know in fact that the times represent US eastern time values, pd.Series.dt.tz_localize can help you express that:

```
ser_no_tz = pd.Series([
    "2024-01-01 00:00:00",
    "2024-01-01 00:01:10",
    "2024-01-01 00:02:42"
], dtype="datetime64[ns]")
ser_et = ser_no_tz.dt.tz_localize("America/New_York")
ser_et
```

```
0    2024-01-01 00:00:00-05:00
1    2024-01-01 00:01:10-05:00
2    2024-01-01 00:02:42-05:00
dtype: datetime64[ns, America/New_York]
```

You can also use pd.Series.dt.tz_convert to translate times into another timezone:

```
ser_pt = ser_et.dt.tz_convert("America/Los_Angeles")
ser_pt
```

```
0    2023-12-31 21:00:00-08:00
1    2023-12-31 21:01:10-08:00
2    2023-12-31 21:02:42-08:00
dtype: datetime64[ns, America/Los_Angeles]
```

You could even set all of your datetime data to midnight of whichever timezone it is in using `pd.Series.dt.normalize`. This can be useful if you don't really care about the time component of your datetimes at all, and just want to treat them as dates, even though pandas does not offer a first-class `DATE` type:

```
ser_pt.dt.normalize()
```

```
0    2023-12-31 00:00:00-08:00
1    2023-12-31 00:00:00-08:00
2    2023-12-31 00:00:00-08:00
dtype: datetime64[ns, America/Los_Angeles]
```

While we have so far pointed out many great features of pandas when working with datetime data, we should also take a look at one of the not-so-great aspects. Back in *Missing value handling*, we talked about how `np.nan` was historically used as a missing value indicator in pandas, even though more modern data types use `pd.NA`. With datetime data types, there is even yet another missing value indicator of `pd.NaT`:

```
ser = pd.Series([
    "2024-01-01",
    None,
    "2024-01-03"
], dtype="datetime64[ns]")
ser
```

```
0    2024-01-01
1           NaT
2    2024-01-03
dtype: datetime64[ns]
```

Again, this difference owes to the history that temporal types were offered before pandas introduced its extension types, and progress to move to one consistent missing value indicator has not fully occurred. Fortunately, functions like `pd.isna` will still correctly identify `pd.NaT` as a missing value:

```
pd.isna(ser)
```

```
0    False
1     True
2    False
dtype: bool
```

There's more...

The historical `ns` precision to pandas limited timestamps to a range that started slightly before 1677-09-21 and would go up to slightly after 2264-04-11. Attempting to assign a datetime value outside of those bounds would raise an `OutOfBoundsDatetime` exception:

```
pd.Series([
    "1500-01-01 00:00:01",
    "2500-01-01 00:00:01",
], dtype="datetime64[ns]")
```

```
OutOfBoundsDatetime: Out of bounds nanosecond timestamp: 1500-01-01 00:00:01,
at position 0
```

Starting with the 3.0 series of pandas, you could specify lower precisions like s, ms, or us to extend your range beyond those windows:

```
pd.Series([
    "1500-01-01 00:00:01",
    "2500-01-01 00:00:01",
], dtype="datetime64[us]")
```

```
0    1500-01-01 00:00:01
1    2500-01-01 00:00:01
dtype: datetime64[us]
```

Temporal types — timedelta

Timedeltas are useful for measuring the *duration* between two points in time. This can be used to measure things like "on average, how much time passed between events X and Y," which can be helpful to monitor and predict the turnaround time of certain processes and/or systems within your organization. Additionally, timedeltas can be used to manipulate your datetimes, making it easy to "add X number of days" or "subtract Y number of seconds" from your datetimes, all without having to dive into the minutiae of how your datetime objects are stored internally.

How to do it

So far, we have introduced each data type by constructing it directly. However, the use cases where you would construct a timedelta pd.Series by hand are exceedingly rare. More commonly, you will come across this type as the result of an expression that subtracts two datetimes from one another:

```
ser = pd.Series([
    "2024-01-01",
    "2024-01-02",
    "2024-01-03"
], dtype="datetime64[ns]")
ser - pd.Timestamp("2023-12-31 12:00:00")
```

```
0    0 days 12:00:00
1    1 days 12:00:00
2    2 days 12:00:00
dtype: timedelta64[ns]
```

Within pandas, there is also the `pd.Timedelta` scalar, which can be used in expressions to add or subtract a duration to datetimes. For instance, the following code shows you how to add 3 days to every datetime in a `pd.Series`:

```
ser + pd.Timedelta("3 days")
```

```
0    2024-01-04
1    2024-01-05
2    2024-01-06
dtype: datetime64[ns]
```

There's more...

While not a common pattern, if you ever needed to manually construct a `pd.Series` of timedelta objects, you could do so using `dtype="timedelta[ns]"`:

```
pd.Series([
    "-1 days",
    "6 hours",
    "42 minutes",
    "12 seconds",
    "8 milliseconds",
    "4 microseconds",
    "300 nanoseconds",
], dtype="timedelta64[ns]")
```

```
0            -1 days +00:00:00
1             0 days 06:00:00
2             0 days 00:42:00
3             0 days 00:00:12
4       0 days 00:00:00.008000
5       0 days 00:00:00.000004
6    0 days 00:00:00.000000300
dtype: timedelta64[ns]
```

What if we tried to create a timedelta of months? Let's see:

```
pd.Series([
    "1 months",
], dtype="timedelta64[ns]")
```

```
ValueError: invalid unit abbreviation: months
```

The reason pandas does not allow this is that timedelta represents a consistently measurable *duration*. While there are always 1,000 nanoseconds in a microsecond, 1,000 microseconds in a millisecond, 1,000 milliseconds in a second, and so on, the number of days in a month is not consistent, ranging from 28-31. Saying two events occurred *one month apart* does not appease the rather strict requirements of a timedelta to measure a finite duration of time passed between two points.

If you need the ability to move dates by the calendar rather than by a finite duration, you can still use the pd.DateOffset object we will introduce in *Chapter 9, Temporal Data Types and Algorithms*. While this does not have an associated data type to introduce in this chapter, the object itself can be a great complement or augmentation of the timedelta type, for analyses that don't strictly think of time as a finite duration.

Temporal PyArrow types

At this point, we have reviewed many of the "first-class" data types built into pandas, while highlighting some rough edges and inconsistencies that plague them. Despite those issues, the types baked into pandas can take you a long way in your data journey.

But there are still cases where the pandas types are not suitable, with a common case being interoperability with databases. Most databases have distinct DATE and DATETIME types, so the fact that pandas only offers a DATETIME type can be disappointing to users fluent in SQL.

Fortunately, the Apache Arrow project defines a true DATE type. Starting in version 2.0, pandas users can start leveraging Arrow types exposed through the PyArrow library.

How to do it

To construct PyArrow types in pandas directly, you will always provide a dtype= argument of the pd.ArrowDtype(XXX) form, replacing XXX with the appropriate PyArrow type. The DATE type in PyArrow is called pa.date32():

```
ser = pd.Series([
    "2024-01-01",
    "2024-01-02",
    "2024-01-03",
], dtype=pd.ArrowDtype(pa.date32()))
ser
```

```
0    2024-01-01
1    2024-01-02
2    2024-01-03
dtype: date32[day][pyarrow]
```

The pa.date32() type can express a wider range of dates without having to toggle the precision:

```
ser = pd.Series([
    "9999-12-29",
```

```
    "9999-12-30",
    "9999-12-31",
], dtype=pd.ArrowDtype(pa.date32()))
ser
```

```
0      9999-12-29
1      9999-12-30
2      9999-12-31
dtype: date32[day][pyarrow]
```

The PyArrow library offers a timestamp type; however, the functionality is nearly identical to the datetime type you have already seen, so I would advise sticking with the datetime type built into pandas.

PyArrow List types

Life would be so simple if every bit of data you came across fit nicely and squarely in a single location of pd.DataFrame, but inevitably you will run into issues where that is not the case. For a second, let's imagine trying to analyze the employees that work at a company:

```
df = pd.DataFrame({
    "name": ["Alice", "Bob", "Janice", "Jim", "Michael"],
    "years_exp": [10, 2, 4, 8, 6],
})
df
```

```
        name        years_exp
0       Alice       10
1       Bob         2
2       Janice      4
3       Jim         8
4       Michael     6
```

This type of data is pretty easy to work with – you could easily add up or take the average number of years that each employee has of experience. But what if we also wanted to know that Bob and Michael reported to Alice while Janice reported to Jim?

Our picturesque view of the world has suddenly come crashing down – how could we possibly express this in pd.DataFrame? If you are coming from a Microsoft Excel or SQL background, you may be tempted to think that you need to create a separate pd.DataFrame that holds the direct reports information. In pandas, we can express this more naturally using the PyArrow pa.list_() data type.

How to do it

When working with a pa.list_() type, you must *parametrize* it with the data type of elements it will contain. In our case, we want our list to contain values like Bob and Janice, so we will parametrize our pa.list_() type with the pa.string() type:

```
ser = pd.Series([
    ["Bob", "Michael"],
    None,
    None,
    ["Janice"],
    None,
], dtype=pd.ArrowDtype(pa.list_(pa.string())))
df["direct_reports"] = ser
df
```

```
     name      years_exp    direct_reports
0    Alice     10           ['Bob' 'Michael']
1    Bob       2            <NA>
2    Janice    4            <NA>
3    Jim       8            ['Janice']
4    Michael   6            <NA>
```

There's more...

When working with a pd.Series that has a PyArrow list type, you can unlock more features of the pd.Series by using the .list accessor. For instance, to see how many items a list contains, you can call ser.list.len():

```
ser.list.len()
```

```
0        2
1     <NA>
2     <NA>
3        1
4     <NA>
dtype: int32[pyarrow]
```

You can access the list item at a given position using the .list[] syntax:

```
ser.list[0]
```

```
0      Bob
1     <NA>
2     <NA>
3   Janice
4     <NA>
dtype: string[pyarrow]
```

There's also a .list.flatten accessor, which could help you identify all of the employees who report to someone:

```
ser.list.flatten()
```

```
0         Bob
1      Michael
2       Janice
dtype: string[pyarrow]
```

PyArrow decimal types

When we looked at the *Floating point types* recipe earlier in this chapter, one of the important things we mentioned was that floating types are *inexact*. Most users of computer software can go their entire lives without knowing this fact, and in many cases, the lack of precision may be an acceptable trade-off to get the performance offered by floating point types. However, in some domains, it is **critical** to have extremely precise computations.

As a simplistic example, let's assume that a movie recommender system used floating point arithmetic to calculate the rating for a given movie as 4.3334 out of 5 stars when it *really* should have been 4.33337. Even if that rounding error was repeated a million times, it probably wouldn't have a largely negative effect on civilization. On the flip side, a financial system that processes billions of transactions per day would find this rounding error to be unacceptable. Over time, that rounding error would accumulate into a rather large number in its own right.

Decimal data types are the solution to these problems. By giving up some performance that you would get with floating point calculations, decimal values allow you to achieve more precise calculations.

How to do it

The `pa.decimal128()` data type requires two arguments that define the *precision* and *scale* of the numbers you wish to represent. The precision dictates how many decimal digits can safely be stored, with the scale representing how many of those decimal digits may appear after a decimal point.

For example, with a *precision* of 5 and a *scale* of 2, you would be able to accurately represent numbers between -999.99 and 999.99, whereas a precision of 5 with a scale of 0 gives you a range of -99999 to 99999. In practice, the precision you choose will be much higher.

Here's an example of how to represent this in a `pd.Series`:

```
pd.Series([
    "123456789.123456789",
    "-987654321.987654321",
    "99999999.9999999999",
], dtype=pd.ArrowDtype(pa.decimal128(19, 10)))
```

```
0       123456789.1234567890
1      -987654321.9876543210
2        99999999.9999999999
dtype: decimal128(19, 10)[pyarrow]
```

Pay special attention to the fact that we provided our data as strings. If we had tried to provide that as floating point data to begin with, we would have immediately seen a loss in precision:

```
pd.Series([
    123456789.123456789,
    -987654321.987654321,
    99999999.9999999999,
], dtype=pd.ArrowDtype(pa.decimal128(19, 10)))
```

```
0        123456789.1234567910
1       -987654321.9876543283
2        100000000.0000000000
dtype: decimal128(19, 10)[pyarrow]
```

This happens because Python itself uses floating point storage for real numbers by default, so the rounding error happens the moment the language runtime tries to interpret the numbers you have provided. Depending on your platform, you may even find that `99999999.9999999999 == 100000000.0` returns `True`. To a human reader, that is obviously not true, but the limits of computer storage prevent the language from being able to discern that.

Python's solution to this issue is the `decimal` module, which ensures rounding errors do not occur:

```
import decimal
decimal.Decimal("99999999.9999999999") == decimal.Decimal("100000000.0")
```

```
False
```

While still giving you proper arithmetic, as follows:

```
decimal.Decimal("99999999.9999999999") + decimal.Decimal("100000000.0")
```

```
Decimal('199999999.9999999999')
```

`decimal.Decimal` objects are also valid arguments when constructing the PyArrow decimal type:

```
pd.Series([
    decimal.Decimal("123456789.123456789"),
    decimal.Decimal("-987654321.987654321"),
    decimal.Decimal("99999999.9999999999"),
], dtype=pd.ArrowDtype(pa.decimal128(19, 10)))
```

```
0        123456789.1234567890
1       -987654321.9876543210
2         99999999.9999999999
dtype: decimal128(19, 10)[pyarrow]
```

There's more...

The `pa.decimal128` data type can only support up to 38 significant decimal digits. If you need more than that, the Arrow ecosystem also provides a `pa.decimal256` data type:

```
ser = pd.Series([
    "12345678912345678912345678912345678.123456789"
], dtype=pd.ArrowDtype(pa.decimal256(76, 10)))
ser
```

```
0    12345678912345678912345678912345678.1234567890
dtype: decimal256(76, 10)[pyarrow]
```

Just be aware that this will consume twice as much memory as the `pa.decimal128` data type, with potentially even slower calculation times.

NumPy type system, the object type, and pitfalls

As mentioned back in the introduction to this chapter, at least in the 2.x and 3.x series, pandas still defaults to types that are sub-optimal for general data analysis. You will undoubtedly come across them in code from peer or online snippets, however, so understanding how they work, their pitfalls, and how to avoid them will be important for years to come.

How to do it

Let's look at the default construction of a `pd.Series` from a sequence of integers:

```
pd.Series([0, 1, 2])
```

```
0    0
1    1
2    2
dtype: int64
```

From this argument, pandas gave us back a `pd.Series` with an `int64` data type. That seems normal, so what is the big deal? Well, let's go ahead and see what happens when you introduce missing values:

```
pd.Series([0, None, 2])
```

```
0    0.0
1    NaN
2    2.0
dtype: float64
```

Huh? We provided integer data but now we got back a floating point type. Surely specifying the `dtype=` argument will help us fix this:

```
pd.Series([0, None, 2], dtype=int)
```

```
TypeError: int() argument must be a string, a bytes-like object or a number,
not 'NoneType'
```

Try as hard as you might, you simply *cannot* mix missing values with the NumPy integer data type, which pandas returns by default. A common solution to this pattern is to start filling missing values with another value like 0 before casting back to an actual integer data type with pd.Series.astype:

```
ser = pd.Series([0, None, 2])
ser.fillna(0).astype(int)
```

```
0    0
1    0
2    2
dtype: int64
```

That solves the problem of getting us to a proper integer type, but it had to change the data to get us there. Whether this matters is a context-dependent issue; some users may be OK with treating missing values as 0 if all they wanted to do was *sum* the column, but that same user might not be happy with the new *count* and *average* that gets produced by that data.

Note the difference between this fillna approach and using the pandas extension types introduced at the start of this chapter:

```
pd.Series([0, None, 2]).fillna(0).astype(int).mean()
```

```
0.6666666666666666
```

```
pd.Series([0, None, 2], dtype=pd.Int64Dtype()).mean()
```

```
1.0
```

Not only do we get different results, but the approach where we do not use dtype=pd.Int64Dtype() takes longer to compute:

```
import timeit
func = lambda: pd.Series([0, None, 2]).fillna(0).astype(int).mean()
timeit.timeit(func, number=10_000)
```

```
0.9819313539992436
```

```
func = lambda: pd.Series([0, None, 2], dtype=pd.Int64Dtype()).mean()
timeit.timeit(func, number=10_000)
```

```
0.6182142379984725
```

This is perhaps not surprising when you consider the number of steps you had to go through to just get integers instead of floats.

When you look at the historical Boolean data type in pandas, things get even stranger. Let's once again start with the seemingly sane base case:

```
pd.Series([True, False])
```

```
0    True
1    False
dtype: bool
```

Let's throw a wrench into things with a missing value:

```
pd.Series([True, False, None])
```

```
0    True
1    False
2    None
dtype: object
```

This is the first time we have seen the object data type. Sparing some technical details, you should trust that the object data type is one of the worst data types to use in pandas. Essentially *anything* goes with an object data type; it completely disallows the type system from enforcing anything about your data. Even though we just want to store True=/=False values where some may be missing, really any valid value can now be placed alongside those values:

```
pd.Series([True, False, None, "one of these things", ["is not like"], ["the
other"]])
```

```
0                   True
1                  False
2                   None
3    one of these things
4         [is not like]
5           [the other]
dtype: object
```

All of this nonsense can be avoided by using pd.BooleanDtype:

```
pd.Series([True, False, None], dtype=pd.BooleanDtype())
```

```
0    True
1    False
2    <NA>
dtype: boolean
```

Another rather unfortunate fact of the default pandas implementation (at least in the 2.x series) is that the object data type is used for strings:

```
pd.Series(["foo", "bar", "baz"])
```

```
0      foo
1      bar
2      baz
dtype: object
```

Once again, there is nothing there that strictly enforces we have string data:

```
ser = pd.Series(["foo", "bar", "baz"])
ser.iloc[2] = 42
ser
```

```
0      foo
1      bar
2       42
dtype: object
```

With pd.StringDtype(), that type of assignment would raise an error:

```
ser = pd.Series(["foo", "bar", "baz"], dtype=pd.StringDtype())
ser.iloc[2] = 42
```

```
TypeError: Cannot set non-string value '42' into a StringArray.
```

There's more...

We have talked at length in this recipe about how the lack of type enforcement with the object data type is a problem. On the flip side, there are some use cases where having that flexibility can be helpful, especially when interacting with Python objects where you cannot make assertions about the data up front:

```
alist = [42, "foo", ["sub", "list"], {"key": "value"}]
ser = pd.Series(alist)
ser
```

```
0                 42
1                foo
2        [sub, list]
3     {'key': 'value'}
dtype: object
```

If you have worked with a tool like Microsoft Excel in the past, the idea that you can put any value anywhere in almost any format may not seem that novel. On the flip side, if your experience is more based on using SQL databases, the idea that you could just load *any* data may be a foreign concept.

In the realm of data processing, there are two major approaches: **extract, transform, load** (ETL) and **extract, load, transform** (ELT). ETL requires you to *transform* your data before you can load it into a data analysis tool, meaning all of the cleansing has to be done upfront in another tool.

The ELT approach allows you to just load the data first and deal with cleaning it up later; the object data type enables you to use the ELT approach in pandas, should you so choose.

With that said, I would generally advise that you strictly use the object data type as a staging data type before transforming it into a more concrete type. By avoiding the object data type, you will achieve much higher performance, have a better understanding of your data, and be able to write cleaner code.

As a final note in this chapter, it is pretty easy to control data types when you work with a pd.Series constructor directly with the dtype= argument. While the pd.DataFrame also has a dtype= argument, it does not allow you to specify types per column, meaning you usually will end up with the historical NumPy data types when creating a pd.DataFrame:

```
df = pd.DataFrame([
    ["foo", 1, 123.45],
    ["bar", 2, 333.33],
    ["baz", 3, 999.99],
], columns=list("abc"))
df
```

```
     a     b    c
0   foo    1   123.45
1   bar    2   333.33
2   baz    3   999.99
```

Checking pd.DataFrame.dtypes will help us confirm this:

```
df.dtypes
```

```
a       object
b        int64
c      float64
dtype: object
```

To get us into using the more desirable pandas extension types, we could either explicitly use the pd.DataFrame.astype method:

```
df.astype({
    "a": pd.StringDtype(),
    "b": pd.Int64Dtype(),
    "c": pd.Float64Dtype(),
}).dtypes
```

```
a       string[python]
b                Int64
c              Float64
dtype: object
```

Or, we could use the pd.DataFrame.convert_dtypes method with dtype_backend="numpy_nullable":

```
df.convert_dtypes(dtype_backend="numpy_nullable").dtypes
```

```
a      string[python]
b               Int64
c             Float64
dtype: object
```

The term numpy_nullable is a bit of a misnomer at this point in the history of pandas, but, as we mentioned back in the introduction, it was the original name for what later became referred to as the pandas extension type system.

Join our community on Discord

Join our community's Discord space for discussions with the authors and other readers:

```
https://packt.link/pandas
```

4

The pandas I/O System

So far, we have been creating our pd.Series and pd.DataFrame objects *inline* with data. While this is helpful for establishing a theoretical foundation, very rarely would a user do this in production code. Instead, users would use the pandas I/O functions to read/write data from/to various formats.

I/O, which is short for **input/output**, generally refers to the process of reading from and writing to common data formats like CSV, Microsoft Excel, JSON, etc. There is, of course, not just one format for data storage, and many of these options represent trade-offs between performance, storage size, third-party integration, accessibility, and/or ubiquity. Some formats assume well-structured, stringently defined data (SQL being arguably the most extreme), whereas other formats can be used to represent semi-structured data that is not restricted to being two-dimensional (JSON being great example).

The fact that pandas can interact with so many of these data formats is one of its greatest strengths, allowing pandas to be the proverbial Swiss army knife of data analysis tools. Whether you are interacting with SQL databases, a set of Microsoft Excel files, HTML web pages, or a REST API endpoint that transmits data via JSON, pandas is up to the task of helping you build a cohesive view of all of your data. For this reason, pandas is considered a popular tool in the domain of ETL.

We are going to cover the following recipes in this chapter:

- CSV – basic reading/writing
- CSV – strategies for reading large files
- Microsoft Excel – basic reading/writing
- Microsoft Excel – finding tables in non-default locations
- Microsoft Excel – hierarchical data
- SQL using SQLAlchemy
- SQL using ADBC
- Apache Parquet
- JSON

- HTML
- Pickle
- Third party I/O libraries

CSV — basic reading/writing

CSV, which stands for *comma-separated values*, is one of the most common formats for data exchange. While there is no official standard that defines what a CSV file is, most developers and users would loosely consider it to be a plain text file, where each line in the file represents a row of data, and within each row, there are *delimiters* between each field to indicate when one record ends and the next begins. The most commonly used *delimiter* is a comma (hence the name *comma-separated values*), but this is not a hard requirement; it is not uncommon to see CSV files that use a pipe (|), tilde (~), or backtick (`) character as the delimiter. If the delimiter character is expected to appear within a given record, usually some type of quoting surrounds the individual record (or all records) to allow a proper interpretation.

For example, let's assume a CSV file uses a pipe separator with the following contents:

```
column1|column2
a|b|c
```

The first row would be read with only two columns of data, whereas the second row would contain three columns of data. Assuming we wanted the records ["a|b", "c"] to appear in the second row, proper quoting would be required:

```
column1|column2
"a|b"|c
```

The above rules are relatively simple and make it easy to write CSV files, but that in turn makes reading CSV files much more difficult. The CSV format provides no metadata (i.e., what delimiter, quoting rule, etc.), nor does it provide any information about the type of data being provided (i.e., what type of data should be located in column X). This puts the onus on CSV readers to figure this all out on their own, which adds performance overhead and can easily lead to a misinterpretation of data. Being a text-based format, CSV is also an inefficient way of storing data compared to binary formats like Apache Parquet. Some of this can be offset by compressing CSV files (at the cost of read/write performance), but generally, CSV rates as one of the worst formats for CPU efficiency, memory usage, and losslessness.

Despite these shortcomings and more, the CSV format has been around for a long time and won't disappear any time soon, so it is beneficial to know how to read and write such files with pandas.

How to do it

Let's start with a simple pd.DataFrame. Building on our knowledge in *Chapter 3, Data Types*, we know that the default types used by pandas are less than ideal, so we are going to use pd.DataFrame. convert_dtypes with the dtype_backend="numpy_nullable" argument to construct this and all of our pd.DataFrame objects going forward.

```
df = pd.DataFrame([
```

```
        ["Paul", "McCartney", 1942],
        ["John", "Lennon", 1940],
        ["Richard", "Starkey", 1940],
        ["George", "Harrison", 1943],
], columns=["first", "last", "birth"])
df = df.convert_dtypes(dtype_backend="numpy_nullable")
df
```

```
        first    last       birth
0       Paul     McCartney  1942
1       John     Lennon     1940
2       Richard  Starkey    1940
3       George   Harrison   1943
```

💡 **Quick tip:** Enhance your coding experience with the **AI Code Explainer** and **Quick Copy** features. Open this book in the next-gen Packt Reader. Click the **Copy** button (**1**) to quickly copy code into your coding environment, or click the **Explain** button (**2**) to get the AI assistant to explain a block of code to you.

```
                                                            Copy      Explain

    function calculate(a, b) {                               1          2
        return {sum: a + b};
    };
```

🔒 **The next-gen Packt Reader** is included for free with the purchase of this book. Unlock it by scanning the QR code below or visiting `https://www.packtpub.com/unlock/9781836205876`.

To write this `pd.DataFrame` out to a CSV file, we can use the `pd.DataFrame.to_csv` method. Typically, the first argument you would provide is a filename, but in this example, we will use the `io.StringIO` object instead. An `io.StringIO` object acts like a file but does not save anything to your disk. Instead, it manages the file contents completely in memory, requiring no cleanup and leaving nothing behind on your filesystem:

```
import io
buf = io.StringIO()
```

```
df.to_csv(buf)
print(buf.getvalue())
```

```
,first,last,birth
0,Paul,McCartney,1942
1,John,Lennon,1940
2,Richard,Starkey,1940
3,George,Harrison,1943
```

Now that we have a "file" with CSV data, we can use the `pd.read_csv` function to read this data back in. However, by default, I/O functions in pandas will use the same default data types that a `pd.DataFrame` constructor would use. To avoid that, we can fortunately still use the `dtype_backend="numpy_nullable"` argument with I/O read functions:

```
buf.seek(0)
pd.read_csv(buf, dtype_backend="numpy_nullable")
```

	Unnamed: 0	first	last	birth
0	0	Paul	McCartney	1942
1	1	John	Lennon	1940
2	2	Richard	Starkey	1940
3	3	George	Harrison	1943

Interestingly, the `pd.read_csv` result does not exactly match the `pd.DataFrame` we started with, as it includes a newly added `Unnamed: 0` column. When you call `pd.DataFrame.to_csv`, it will write out both your row index and columns to the CSV file. The CSV format does not allow you to store any extra metadata to indicate which columns in the CSV file should map to the row index versus those that should represent a column in the `pd.DataFrame`, so `pd.read_csv` assumes everything to be a column.

You can rectify this situation by letting `pd.read_csv` know that the first column of data in the CSV file should form the row index with an `index_col=0` argument:

```
buf.seek(0)
pd.read_csv(buf, dtype_backend="numpy_nullable", index_col=0)
```

	first	last	birth
0	Paul	McCartney	1942
1	John	Lennon	1940
2	Richard	Starkey	1940
3	George	Harrison	1943

Alternatively, you could avoid writing the index in the first place with the `index=False` argument of `pd.DataFrame.to_csv`:

```
buf = io.StringIO()
df.to_csv(buf, index=False)
print(buf.getvalue())
```

```
first,last,birth
Paul,McCartney,1942
John,Lennon,1940
Richard,Starkey,1940
George,Harrison,1943
```

There's more...

As mentioned back at the beginning of this section, CSV files use quoting to prevent any confusion between the appearance of the *delimiter* within a field and its intended use – to indicate the start of a new record. Fortunately, pandas handles this rather sanely by default, which we can see with some new sample data:

```
df = pd.DataFrame([
    ["McCartney, Paul", 1942],
    ["Lennon, John", 1940],
    ["Starkey, Richard", 1940],
    ["Harrison, George", 1943],
], columns=["name", "birth"])
df = df.convert_dtypes(dtype_backend="numpy_nullable")
df
```

```
    name              birth
0   McCartney, Paul   1942
1   Lennon, John      1940
2   Starkey, Richard  1940
3   Harrison, George  1943
```

Now that we just have a name column that contains a comma, you can see that pandas quotes the field to indicate that the usage of a comma is part of the data itself and not a new record:

```
buf = io.StringIO()
df.to_csv(buf, index=False)
print(buf.getvalue())
```

```
name,birth
"McCartney, Paul",1942
"Lennon, John",1940
"Starkey, Richard",1940
"Harrison, George",1943
```

We could have alternatively decided upon the usage of a different *delimiter*, which can be toggled with the sep= argument:

```
buf = io.StringIO()
df.to_csv(buf, index=False, sep="|")
print(buf.getvalue())
```

```
name|birth
McCartney, Paul|1942
Lennon, John|1940
Starkey, Richard|1940
Harrison, George|1943
```

We also mentioned that, while CSV files are naturally plain text, you can also compress them to save storage space. The easiest way to do this is to provide a filename argument with a common compression file extension, i.e., by saying `df.to_csv("data.csv.zip")`. For more explicit control, you can use the `compression=` argument.

To see this in action, let's work with a larger `pd.DataFrame`:

```
df = pd.DataFrame({
    "col1": ["a"] * 1_000,
    "col2": ["b"] * 1_000,
    "col3": ["c"] * 1_000,
})
df = df.convert_dtypes(dtype_backend="numpy_nullable")
df.head()
```

	col1	col2	col3
0	a	b	c
1	a	b	c
2	a	b	c
3	a	b	c
4	a	b	c

Take note of the number of bytes used to write this out as a plain text CSV file:

```
buf = io.StringIO()
df.to_csv(buf, index=False)
len(buf.getvalue())
```

```
6015
```

Using `compression="gzip"`, we can produce a file with far less storage:

```
buf = io.BytesIO()
df.to_csv(buf, index=False, compression="gzip")
len(buf.getvalue())
```

```
69
```

The trade-off here is that while compressed files require less disk storage, they require more work from the CPU to compress or decompress the file contents.

CSV – strategies for reading large files

Handling large CSV files can be challenging, especially when they exhaust the memory of your computer. In many real-world data analysis scenarios, you might encounter datasets that are too large to be processed in a single-read operation. This can lead to performance bottlenecks and MemoryError exceptions, making it difficult to proceed with your analysis. However, fear not! There are quite a few levers you can pull to more efficiently try and process files.

In this recipe, we will show you how you can use pandas to peek at parts of your CSV file to understand what data types are being inferred. With that understanding, we can instruct pd.read_csv to use more efficient data types, yielding far more efficient memory usage.

How to do it

For this example, we will look at the *diamonds* dataset. This dataset is not actually all that big for modern computers, but let's pretend that the file is a lot bigger than it is, or that the memory on our machine is limited to the point where a normal read_csv call would yield a MemoryError.

To start, we will look at the first 1,000 rows from the dataset to get an idea of what is in the file via nrows=1_000.

```
df = pd.read_csv("data/diamonds.csv", dtype_backend="numpy_nullable",
nrows=1_000)
df
```

	carat	cut	color	clarity	depth	table	price	x	y	z
0	0.23	Ideal	E	SI2	61.5	55.0	326	3.95	3.98	2.43
1	0.21	Premium	E	SI1	59.8	61.0	326	3.89	3.84	2.31
2	0.23	Good	E	VS1	56.9	65.0	327	4.05	4.07	2.31
3	0.29	Premium	I	VS2	62.4	58.0	334	4.2	4.23	2.63
4	0.31	Good	J	SI2	63.3	58.0	335	4.34	4.35	2.75
...	...	...	...	...	...	...	...	...	...	...
995	0.54	Ideal	D	VVS2	61.4	52.0	2897	5.3	5.34	3.26
996	0.72	Ideal	E	SI1	62.5	55.0	2897	5.69	5.74	3.57
997	0.72	Good	F	VS1	59.4	61.0	2897	5.82	5.89	3.48
998	0.74	Premium	D	VS2	61.8	58.0	2897	5.81	5.77	3.58
999	1.12	Premium	J	SI2	60.6	59.0	2898	6.68	6.61	4.03

1000 rows × 10 columns

The pd.DataFrame.info method should give us an idea of how much memory this subset uses:

```
df.info()
```

```
<class 'pandas.core.frame.DataFrame'>
RangeIndex: 1000 entries, 0 to 999
Data columns (total 10 columns):
```

```
 #    Column   Non-Null Count  Dtype
---   ------   --------------  -----
 0    carat    1000 non-null   Float64
 1    cut      1000 non-null   string
 2    color    1000 non-null   string
 3    clarity  1000 non-null   string
 4    depth    1000 non-null   Float64
 5    table    1000 non-null   Float64
 6    price    1000 non-null   Int64
 7    x        1000 non-null   Float64
 8    y        1000 non-null   Float64
 9    z        1000 non-null   Float64
dtypes: Float64(6), Int64(1), string(3)
memory usage: 85.1 KB
```

The exact memory usage you see may depend on your version of pandas and operating system, but let's assume that the pd.DataFrame we are using requires around 85 KB of memory. If we had 1 billion rows instead of just 1,000, that would require 85 GB of memory just to store this pd.DataFrame.

So how can we fix this situation? For starters, it is worth looking more closely at the data types that have been inferred. The price column may be one that immediately catches our attention; this was inferred to be a pd.Int64Dtype(), but chances are that we don't need 64 bits to store this information. Summary statistics will be explored in more detail in *Chapter 5, Algorithms and How to Apply Them* but for now, let's just take a look at pd.Series.describe to see what pandas can tell us about this column:

```
df["price"].describe()
```

```
count        1000.0
mean        2476.54
std       839.57562
min          326.0
25%         2777.0
50%         2818.0
75%         2856.0
max         2898.0
Name: price, dtype: Float64
```

The minimum value is 326 and the maximum is 2,898. Those values can both safely fit into pd.Int16Dtype(), which would represent good memory savings compared to pd.Int64Dtype().

Let's also take a look at some of the floating point types, starting with the *carat*:

```
df["carat"].describe()
```

```
count        1000.0
mean        0.68928
```

```
std        0.195291
min            0.2
25%            0.7
50%            0.71
75%            0.79
max            1.27
Name: carat, dtype: Float64
```

The values range from 0.2 to 1.27, and unless we expect to perform calculations with many decimal points, the 6 to 9 digits of decimal precision that a 32-bit floating point data type provides should be good enough to use here.

For this recipe, we are going to assume that 32-bit floating point types can be used across all of the other floating point types as well. One way to tell pd.read_csv that we want to use smaller data types would be to use the dtype= parameter, with a dictionary mapping column names to the desired types. Since our dtype= parameter will cover all of the columns, we can also drop dtype_backend="numpy_nullable", as it would be superfluous:

```python
df2 = pd.read_csv(
    "data/diamonds.csv",
    nrows=1_000,
    dtype={
        "carat": pd.Float32Dtype(),
        "cut": pd.StringDtype(),
        "color": pd.StringDtype(),
        "clarity": pd.StringDtype(),
        "depth": pd.Float32Dtype(),
        "table": pd.Float32Dtype(),
        "price": pd.Int16Dtype(),
        "x": pd.Float32Dtype(),
        "y": pd.Float32Dtype(),
        "z": pd.Float32Dtype(),
    }
)
df2.info()
```

```
<class 'pandas.core.frame.DataFrame'>
RangeIndex: 1000 entries, 0 to 999
Data columns (total 10 columns):
 #   Column  Non-Null Count  Dtype
---  ------  --------------  -----
 0   carat   1000 non-null   Float32
 1   cut     1000 non-null   string
```

```
2    color    1000 non-null    string
3    clarity  1000 non-null    string
4    depth    1000 non-null    Float32
5    table    1000 non-null    Float32
6    price    1000 non-null    Int16
7    x        1000 non-null    Float32
8    y        1000 non-null    Float32
9    z        1000 non-null    Float32
dtypes: Float32(6), Int16(1), string(3)
memory usage: 55.8 KB
```

These steps alone will probably yield a memory usage in the ballpark of 55 KB, which is not a bad reduction from the 85 KB we started with! For added safety, we can use the pd.DataFrame.describe() method to get summary statistics and ensure that the two pd.DataFrame objects are similar. If the numbers are the same for both pd.DataFrame objects, it is a good sign that our conversions did not materially change our data:

```
df.describe()
```

	carat	depth	table	price	x	y	z
count	1000.0	1000.0	1000.0	1000.0	1000.0	1000.0	1000.0
mean	0.68928	61.7228	57.7347	2476.54	5.60594	5.59918	3.45753
std	0.195291	1.758879	2.467946	839.57562	0.625173	0.611974	0.389819
min	0.2	53.0	52.0	326.0	3.79	3.75	2.27
25%	0.7	60.9	56.0	2777.0	5.64	5.63	3.45
50%	0.71	61.8	57.0	2818.0	5.77	5.76	3.55
75%	0.79	62.6	59.0	2856.0	5.92	5.91	3.64
max	1.27	69.5	70.0	2898.0	7.12	7.05	4.33

```
df2.describe()
```

	carat	depth	table	price	x	y	z
count	1000.0	1000.0	1000.0	1000.0	1000.0	1000.0	1000.0
mean	0.68928	61.722801	57.734699	2476.54	5.60594	5.59918	3.45753
std	0.195291	1.758879	2.467946	839.57562	0.625173	0.611974	0.389819
min	0.2	53.0	52.0	326.0	3.79	3.75	2.27
25%	0.7	60.900002	56.0	2777.0	5.64	5.63	3.45
50%	0.71	61.799999	57.0	2818.0	5.77	5.76	3.55
75%	0.79	62.599998	59.0	2856.0	5.92	5.91	3.64
max	1.27	69.5	70.0	2898.0	7.12	7.05	4.33

So far, things are looking good, but we can still do better. For starters, it looks like the cut column has a relatively small amount of unique values:

```
df2["cut"].unique()
```

```
<StringArray>
['Ideal', 'Premium', 'Good', 'Very Good', 'Fair']
Length: 5, dtype: string
```

The same can be said about the color column:

```
df2["color"].unique()
```

```
<StringArray>
['E', 'I', 'J', 'H', 'F', 'G', 'D']
Length: 7, dtype: string
```

As well as the clarity column:

```
df2["clarity"].unique()
```

```
<StringArray>
['SI2', 'SI1', 'VS1', 'VS2', 'VVS2', 'VVS1', 'I1', 'IF']
Length: 8, dtype: string
```

Out of 1,000 rows sampled, there are only 5 distinct cut values, 7 distinct color values, and 8 distinct clarity values. We consider these columns to have a *low cardinality*, i.e., the number of distinct values is very low relative to the number of rows.

This makes these columns a perfect candidate for the use of categorical types. However, I would advise against using pd.CategoricalDtype() as an argument to dtype=, as by default it uses np.nan as a missing value indicator (for a refresher on this caveat, you may want to revisit the *Categorical types* recipe back in *Chapter 3*, *Data Types*). Instead, the best approach to convert your strings to categorical types is to first read in your columns as pd.StringDtype(), and then use pd.DataFrame.astype on the appropriate column(s):

```
df3 = pd.read_csv(
    "data/diamonds.csv",
    nrows=1_000,
    dtype={
        "carat": pd.Float32Dtype(),
        "cut": pd.StringDtype(),
        "color": pd.StringDtype(),
        "clarity": pd.StringDtype(),
        "depth": pd.Float32Dtype(),
        "table": pd.Float32Dtype(),
        "price": pd.Int16Dtype(),
```

```
        "x": pd.Float32Dtype(),
        "y": pd.Float32Dtype(),
        "z": pd.Float32Dtype(),
    }
)
cat_cols = ["cut", "color", "clarity"]
df3[cat_cols] = df3[cat_cols].astype(pd.CategoricalDtype())
df3.info()
```

```
<class 'pandas.core.frame.DataFrame'>
RangeIndex: 1000 entries, 0 to 999
Data columns (total 10 columns):
 #   Column   Non-Null Count  Dtype
---  ------   --------------  -----
 0   carat    1000 non-null   Float32
 1   cut      1000 non-null   category
 2   color    1000 non-null   category
 3   clarity  1000 non-null   category
 4   depth    1000 non-null   Float32
 5   table    1000 non-null   Float32
 6   price    1000 non-null   Int16
 7   x        1000 non-null   Float32
 8   y        1000 non-null   Float32
 9   z        1000 non-null   Float32
dtypes: Float32(6), Int16(1), category(3)
memory usage: 36.2 KB
```

To get even more savings, we may decide that there are columns in our CSV file that are just not worth reading at all. To allow pandas to skip this data and save even more memory, you can use the usecols= parameter:

```
dtypes = {  # does not include x, y, or z
    "carat": pd.Float32Dtype(),
    "cut": pd.StringDtype(),
    "color": pd.StringDtype(),
    "clarity": pd.StringDtype(),
    "depth": pd.Float32Dtype(),
    "table": pd.Float32Dtype(),
    "price": pd.Int16Dtype(),
}
df4 = pd.read_csv(
    "data/diamonds.csv",
    nrows=1_000,
```

```
        dtype=dtypes,
        usecols=dtypes.keys(),
    )
    cat_cols = ["cut", "color", "clarity"]
    df4[cat_cols] = df4[cat_cols].astype(pd.CategoricalDtype())
    df4.info()
```

```
<class 'pandas.core.frame.DataFrame'>
RangeIndex: 1000 entries, 0 to 999
Data columns (total 7 columns):
 #   Column   Non-Null Count  Dtype
---  ------   --------------  -----
 0   carat    1000 non-null   Float32
 1   cut      1000 non-null   category
 2   color    1000 non-null   category
 3   clarity  1000 non-null   category
 4   depth    1000 non-null   Float32
 5   table    1000 non-null   Float32
 6   price    1000 non-null   Int16
dtypes: Float32(3), Int16(1), category(3)
memory usage: 21.5 KB
```

If the preceding steps are not sufficient to create a small enough pd.DataFrame, you might still be in luck. If you can process chunks of data at a time and do not need all of it in memory, you can use the chunksize= parameter to control the size of the chunks you would like to read from a file:

```
dtypes = {    # does not include x, y, or z
    "carat": pd.Float32Dtype(),
    "cut": pd.StringDtype(),
    "color": pd.StringDtype(),
    "clarity": pd.StringDtype(),
    "depth": pd.Float32Dtype(),
    "table": pd.Float32Dtype(),
    "price": pd.Int16Dtype(),
}
df_iter = pd.read_csv(
    "data/diamonds.csv",
    nrows=1_000,
    dtype=dtypes,
    usecols=dtypes.keys(),
    chunksize=200
)
```

```
for df in df_iter:
    cat_cols = ["cut", "color", "clarity"]
    df[cat_cols] = df[cat_cols].astype(pd.CategoricalDtype())
    print(f"processed chunk of shape {df.shape}")
```

```
processed chunk of shape (200, 7)
processed chunk of shape (200, 7)
processed chunk of shape (200, 7)
processed chunk of shape (200, 7)
processed chunk of shape (200, 7)
```

There's more...

The usecols parameter introduced here can also accept a callable that, when evaluated against each column label encountered, should return True if the column should be read and False if it should be skipped. If we only wanted to read the carat, cut, color, and clarity columns, that might look something like:

```
def startswith_c(column_name: str) -> bool:
    return column_name.startswith("c")


pd.read_csv(
    "data/diamonds.csv",
    dtype_backend="numpy_nullable",
    usecols=startswith_c,
)
```

	carat	cut	color	clarity
0	0.23	Ideal	E	SI2
1	0.21	Premium	E	SI1
2	0.23	Good	E	VS1
3	0.29	Premium	I	VS2
4	0.31	Good	J	SI2
...	...	...	...	...
53935	0.72	Ideal	D	SI1
53936	0.72	Good	D	SI1
53937	0.7	Very Good	D	SI1
53938	0.86	Premium	H	SI2
53939	0.75	Ideal	D	SI2

53940 rows × 4 columns

Microsoft Excel — basic reading/writing

Microsoft Excel is an extremely popular tool for data analysis, given its ease of use and ubiquity. Microsoft Excel provides a rather powerful toolkit that can help to cleanse, transform, store, and visualize data, all without requiring any knowledge of programming languages. Many successful analysts may consider it to be the *only* tool they will ever need. Despite this, Microsoft Excel really struggles with performance and scalability and, when used as a storage medium, may even materially change your data in unexpected ways.

If you have used Microsoft Excel before and are now picking up pandas, you will find that pandas works as a complementary tool. With pandas, you will give up the point-and-click usability of Microsoft Excel, but you will easily unlock performance that takes you far beyond the limits of Microsoft Excel.

Before we jump into this recipe, it's worth noting that Microsoft Excel support is not shipped as part of pandas, so you will need to install third-party package(s) for these recipes to work. While it is not the only choice, users are encouraged to opt for installing openpyxl, as it works very well to read and write all of the various Microsoft Excel formats. If you do not have it already, openpyxl can be installed via:

```
python -m pip install openpyxl
```

How to do it

Let's again start with a simple pd.DataFrame:

```
df = pd.DataFrame([
    ["Paul", "McCartney", 1942],
    ["John", "Lennon", 1940],
    ["Richard", "Starkey", 1940],
    ["George", "Harrison", 1943],
], columns=["first", "last", "birth"])
df = df.convert_dtypes(dtype_backend="numpy_nullable")
df
```

```
     first     last       birth
0    Paul      McCartney  1942
1    John      Lennon     1940
2    Richard   Starkey    1940
3    George    Harrison   1943
```

You can use the pd.DataFrame.to_excel method to write this to a file, with the first argument typically being a filename like myfile.xlsx, but here, we will again use io.BytesIO, which acts like a file but stores binary data in memory instead of on disk:

```
import io
buf = io.BytesIO()
df.to_excel(buf)
```

For reading, reach for the `pd.read_excel` function. We will continue to use `dtype_backend="numpy_`
`nullable"` to prevent the default type inference that pandas performs:

```
buf.seek(0)
pd.read_excel(buf, dtype_backend="numpy_nullable")
```

```
   Unnamed: 0   first    last       birth
0  0            Paul     McCartney  1942
1  1            John     Lennon     1940
2  2            Richard  Starkey    1940
3  3            George   Harrison   1943
```

Many of the function parameters are shared with CSV. To get rid of the `Unnamed: 0` column above, we
can either specify the `index_col=` argument:

```
buf.seek(0)
pd.read_excel(buf, dtype_backend="numpy_nullable", index_col=0)
```

```
   first    last       birth
0  Paul     McCartney  1942
1  John     Lennon     1940
2  Richard  Starkey    1940
3  George   Harrison   1943
```

Or choose to not write the index in the first place:

```
buf = io.BytesIO()
df.to_excel(buf, index=False)
buf.seek(0)
pd.read_excel(buf, dtype_backend="numpy_nullable")
```

```
   first    last       birth
0  Paul     McCartney  1942
1  John     Lennon     1940
2  Richard  Starkey    1940
3  George   Harrison   1943
```

Data types can be controlled with the dtype= argument:

```
buf.seek(0)
dtypes = {
    "first": pd.StringDtype(),
    "last": pd.StringDtype(),
    "birth": pd.Int16Dtype(),
}
df = pd.read_excel(buf, dtype=dtypes)
df.dtypes
```

```
first      string[python]
last       string[python]
birth               Int16
dtype: object
```

Microsoft Excel — finding tables in non-default locations

In the previous recipe, Microsoft Excel – basic reading/writing, we used the Microsoft Excel I/O functions without thinking about *where* within the worksheet our data was. By default, pandas will read from / write to the first cell on the first sheet of data, but it is not uncommon to receive Microsoft Excel files where the data you want to read is located elsewhere within the document.

For this example, we have a Microsoft Excel workbook where the very first tab, **Sheet1**, is used as a cover sheet:

Figure 4.1: Workbook where Sheet1 contains no useful data

The second sheet is where we have useful information:

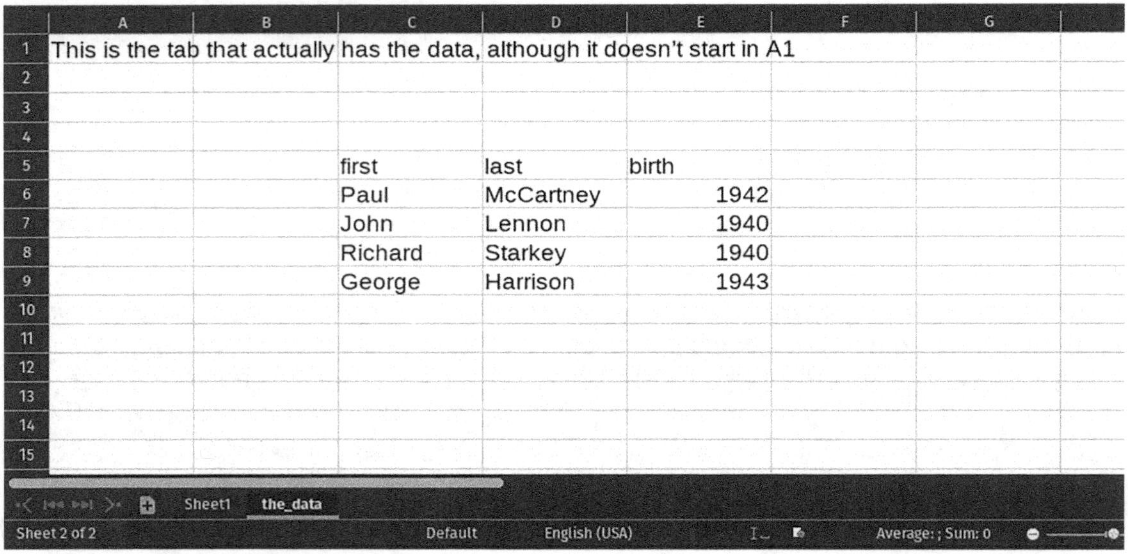

	A	B	C	D	E	F	G
1	This is the tab that actually has the data, although it doesn't start in A1						
2							
3							
4							
5			first	last	birth		
6			Paul	McCartney	1942		
7			John	Lennon	1940		
8			Richard	Starkey	1940		
9			George	Harrison	1943		
10							
11							
12							
13							
14							
15							

Sheet1   the_data

Sheet 2 of 2   Default   English (USA)   Average: ; Sum: 0

Figure 4.2: Workbook where another sheet has relevant data

How to do it

To still be able to read this data, you can use a combination of the sheet_name=, skiprows=, and usecols= arguments to pd.read_excel:

```python
pd.read_excel(
    "data/beatles.xlsx",
    dtype_backend="numpy_nullable",
    sheet_name="the_data",
    skiprows=4,
    usecols="C:E",
)
```

```
      first      last    birth
0     Paul    McCartney   1942
1     John    Lennon      1940
2     Richard Starkey     1940
3     George  Harrison    1943
```

By passing sheet_name="the_data", the pd.read_excel function is able to pinpoint the specific sheet within the Microsoft Excel file to start looking for data. Alternatively, we could have used sheet_name=1 to search by tab position. After locating the correct sheet, pandas looks at the skiprows= argument and knows to ignore rows 1–4 on the worksheet. It then looks at the usecols= argument to select only columns C–E.

There's more...

Instead of usecols="C:E", we could have also provided the labels we wanted:

```python
pd.read_excel(
    "data/beatles.xlsx",
    dtype_backend="numpy_nullable",
    sheet_name="the_data",
    skiprows=4,
    usecols=["first", "last", "birth"],
)
```

```
     first    last      birth
0    Paul     McCartney 1942
1    John     Lennon    1940
2    Richard  Starkey   1940
3    George   Harrison  1943
```

Passing such an argument to usecols= was a requirement when working with the CSV format to select particular columns from a file. However, pandas provides special behavior when reading Microsoft Excel files to allow strings like "C:E" or "C,D,E" to refer to columns.

Microsoft Excel — hierarchical data

One of the major tasks with data analysis is to take very detailed information and aggregate it into a summary that is easy to digest. Rather than having to sift through thousands of orders, most executives at a company just want to know, "What have my sales looked like in the last X quarters?"

With Microsoft Excel, users will commonly summarize this information in a view like the one shown in *Figure 4.3*, which represents a hierarchy of Region/Sub-Region along the rows and Year/Quarter along the columns:

	A	B	C	D	E	F	G
1		Year	2024		2025		
2		Quarter	Q1	Q2	Q1	Q2	
3	Region	Sub-Region					
4		East	1	2	4	8	
5	America	West	16	32	64	128	
6		South	256	512	1024	4096	
7	Europe	West	8192	16384	32768	65536	
8		East	131072	262144	524288	1048576	
9							
10							
11							
12							
13							

Figure 4.3: Workbook with hierarchical data – sales by Region and Quarter

While this summary does not seem too far-fetched, many analysis tools struggle to properly present this type of information. Taking a traditional SQL database as an example, there is no direct way to represent this Year/Quarter hierarchy in a table – your only option would be to concatenate all of the hierarchy fields together and produce columns like 2024/Q1, 2024/Q2, 2025/Q1 and 2025/Q2. While that makes it easy to select any individual column, you lose the ability to easily select things like "all of 2024 sales" without additional effort.

Fortunately, pandas can handle this a lot more sanely than a SQL database can, directly supporting such hierarchical relationships in both the row and column index. If you recall *Chapter 2, Selection and Assignment*, we introduced the pd.MultiIndex; being able to maintain those relationships allows users to efficiently select from any and all levels of the hierarchies.

How to do it

Upon closer inspection of *Figure 4.3*, you will see that rows 1 and 2 contain the labels Year and Quarter, which can form the levels of the pd.MultiIndex that we want in the columns of our pd.DataFrame. Microsoft Excel uses 1-based numbering of each row, so rows [1, 2] translated to Python would actually be [0, 1]; we will use this as our header= argument to establish that we want the first two rows to form our column pd.MultiIndex.

Switching our focus to columns A and B in Microsoft Excel, we can now see the labels Region and Sub-Region, which will help us shape the pd.MultiIndex in our rows. Back in the *CSV – basic reading/writing* section, we introduced the index_col= argument, which can be used to tell pandas which column(s) of data should actually be used to generate the row index. Columns A and B from the Microsoft Excel file represent the first and second columns, so we can once again use [0, 1] to let pandas know our intent:

```
df = pd.read_excel(
    "data/hierarchical.xlsx",
    dtype_backend="numpy_nullable",
    index_col=[0, 1],
    header=[0, 1],
)
df
```

	Year	2024		2025	
	Quarter	Q1	Q2	Q1	Q2
Region	Sub-Region				
America	East	1	2	4	8
	West	16	32	64	128
	South	256	512	1024	4096
Europe	West	8192	16384	32768	65536
	East	131072	262144	524288	1048576

Voila! We have successfully read in the data and maintained the hierarchical nature of the rows and columns, which lets us use all of the native pandas functionality to select from this data, and even answer questions like, "What does the Q2 performance look like year over year for every East Sub-Region?"

```
df.loc[(slice(None), "East"), (slice(None), "Q2")]
```

```
                Year       2024      2025
                Quarter    Q2        Q2
Region   Sub-Region
America  East       2         8
Europe   East       262144    1048576
```

SQL using SQLAlchemy

The pandas library provides robust capabilities for interacting with SQL databases, allowing you to perform data analysis directly on data stored in relational databases.

There are, of course, countless databases that exist (and more are coming!), each with its own features, authentication schemes, dialects, and quirks. To interact with them, pandas relies on another great Python library, SQLAlchemy, which at its core acts as a bridge between Python and the database world. In theory, pandas can work with any database that SQLAlchemy can connect to.

To get started, you should first install SQLAlchemy into your environment:

```
python -m pip install sqlalchemy
```

SQLAlchemy supports all major databases, like MySQL, PostgreSQL, MS SQL Server, etc., but setting up and properly configuring those databases is an effort in its own right, which cannot be covered within the scope of this book. To make things as simple as possible, we will focus on using SQLite as our database, as it requires no setup and can operate entirely within memory on your computer. Once you feel comfortable experimenting with SQLite, you only need to change the credentials you use to point to your target database; otherwise, all of the functionality remains the same.

How to do it

The first thing we need to do is create a SQLAlchemy engine, using `sa.create_engine`. The argument to this function is a URL and will be dependent upon the database you are trying to connect to (see the SQLAlchemy docs for more info). For these examples, we are going to use SQLite in memory:

```
import sqlalchemy as sa
engine = sa.create_engine("sqlite:///:memory:")
```

With the `pd.DataFrame.to_sql` method, you can take an existing `pd.DataFrame` and write it to a database table. The first argument is the name of the table you would like to create, with the second argument being an engine/connectable:

```
df = pd.DataFrame([
    ["dog", 4],
    ["cat", 4],
], columns=["animal", "num_legs"])
df = df.convert_dtypes(dtype_backend="numpy_nullable")

df.to_sql("table_name", engine, index=False)
```

```
2
```

The `pd.read_sql` function can be used to go in the opposite direction and read from a database table:

```
pd.read_sql("table_name", engine, dtype_backend="numpy_nullable")
```

```
   animal    num_legs
0  dog       4
1  cat       4
```

Alternatively, if you wanted something different than just a copy of the table, you could pass a SQL query to `pd.read_sql`:

```
pd.read_sql(
    "SELECT SUM(num_legs) AS total_legs FROM table_name",
    engine,
    dtype_backend="numpy_nullable"
)
```

```
   total_legs
0  8
```

When a table already exists in a database, trying to write to the same table again will raise an error. You can pass `if_exists="replace"` to override this behavior and replace the table:

```
df = pd.DataFrame([
    ["dog", 4],
    ["cat", 4],
    ["human", 2],
], columns=["animal", "num_legs"])
df = df.convert_dtypes(dtype_backend="numpy_nullable")
df.to_sql("table_name", engine, index=False, if_exists="replace")
```

```
3
```

You can also use `if_exists="append"` to add data to a table:

```
new_data = pd.DataFrame([["centipede", 100]], columns=["animal", "num_legs"])
new_data.to_sql("table_name", engine, index=False, if_exists="append")
pd.read_sql("table_name", engine, dtype_backend="numpy_nullable")
```

```
     animal       num_legs
0    dog          4
1    cat          4
2    human        2
3    centipede    100
```

The bulk of the heavy lifting is done behind the scenes by the SQLAlchemy engine, which is construct-ed using a URL of the form `dialect+driver://username:password@host:port/database`. Not all of the fields in that URL are required – the string will depend largely on the database you are using and how it is configured.

In our specific example, `sa.create_engine("sqlite:///:memory:")` creates and connects to a SQLite database in the memory space of our computer. This feature is specific to SQLite; instead of `:memory:`, we could have also passed a path to a file on our computer like `sa.create_engine("sqlite:///tmp/adatabase.sql")`.

For more information on SQLAlchemy URLs and to get an idea of drivers to pair with other databases, see the SQLAlchemy Backend-Specific URLs documentation.

SQL using ADBC

While using SQLAlchemy to connect to databases is a viable option and has helped users of pandas for many years, a new technology has emerged out of the Apache Arrow project that can help scale your SQL interactions even further. This new technology is called **Arrow Database Connectivity**, or **ADBC** for short. Starting in version 2.2, pandas added support for using ADBC drivers to interact with databases.

Using ADBC will offer better performance and type safety when interacting with SQL databases than the aforementioned SQLAlchemy-based approach can. The trade-off is that SQLAlchemy has support for far more databases, so depending on your database, it may be the only option. ADBC maintains a record of its Driver Implementation Status; I would advise looking there first for a stable driver implementation for the database you are using before falling back on SQLAlchemy.

Much like in the previous section, we will use SQLite for our database, given its ease of use to set up and configure. Make sure to install the appropriate ADBC Python package for SQLite:

```
python -m pip install adbc-driver-sqlite
```

How to do it

Let's start by importing the dbapi object from our SQLite ADBC driver and creating some sample data:

```
from adbc_driver_sqlite import dbapi
df = pd.DataFrame([
    ["dog", 4],
    ["cat", 4],
    ["human", 2],
], columns=["animal", "num_legs"])
df = df.convert_dtypes(dtype_backend="numpy_nullable")
df
```

```
      animal    num_legs
0     dog       4
1     cat       4
2     human     2
```

The term dbapi is taken from the Python Database API Specification defined in PEP-249, which standardizes how Python modules and libraries should be used to interact with databases. Calling the .connect method with credentials is the standardized way to open up a database connection in Python. Once again, we will use an in-memory SQLite application via dbapi.connect("file::memory:").

By using the with ... as: syntax to use a context manager in Python, we can connect to a database and assign it to a variable, letting Python automatically clean up the connection when the block is finished. While the connection is open within our block, we can use pd.DataFrame.to_sql / pd.read_sql to write to and read from the database, respectively:

```
with dbapi.connect("file::memory:") as conn:
    df.to_sql("table_name", conn, index=False, if_exists="replace")
    df = pd.read_sql(
        "SELECT * FROM table_name",
        conn,
        dtype_backend="numpy_nullable",
    )
df
```

```
      animal    num_legs
0     dog       4
1     cat       4
2     human     2
```

For smaller datasets, you may not see much of a difference, but the performance gains of ADBC will be drastic with larger datasets. Let's compare the time to write a 10,000 by 10 pd.DataFrame using SQLAlchemy:

```python
import timeit
import sqlalchemy as sa

np.random.seed(42)
df = pd.DataFrame(
    np.random.randn(10_000, 10),
    columns=list("abcdefghij")
)

with sa.create_engine("sqlite:///:memory:").connect() as conn:
    func = lambda: df.to_sql("test_table", conn, if_exists="replace")
    print(timeit.timeit(func, number=100))
```

```
4.898935955003253
```

To equivalent code using ADBC:

```python
from adbc_driver_sqlite import dbapi

with dbapi.connect("file::memory:") as conn:
    func = lambda: df.to_sql("test_table", conn, if_exists="replace")
    print(timeit.timeit(func, number=100))
```

```
0.7935214300014195
```

Your results will vary, depending on your data and database, but generally, ADBC should perform much faster.

There's more...

To understand what ADBC does and why it matters, it is first worth a quick history lesson in database standards and how they have evolved. Back in the 1990s, the **Open Database Connectivity (ODBC)** and **Java Database Connectivity (JDBC)** standards were introduced, which helped standardize how different clients could talk to various databases. Before the introduction of these standards, if you developed an application that needed to work with two or more different databases, your application would have to speak the exact language that each database understood to interact with it.

Imagine then that this application wanted to just get a listing of tables available in each database. A PostgreSQL database stores this information in a table called pg_catalog.pg_tables, whereas a SQLite database stores this in a sqlite_schema table where type='table'. The application would need to be developed with this particular information, and then it would need to be re-released every time a database changed how it stored this information, or if an application wanted to support a new database.

With a standard like ODBC, the application instead just needs to communicate with a driver, letting the driver know that it wants all of the tables in the system. This shifts the onus of properly interacting with a database from the application itself to the driver, giving the application a layer of abstraction. As new databases or versions are released, the application itself no longer needs to change; it simply works with a new ODBC/JDBC driver and continues to work. SQLAlchemy, in fact, is just like this theoretical application; it interacts with databases through either ODBC/JDBC drivers, rather than trying to manage the endless array of database interactions on its own.

While these standards are fantastic for many purposes, it is worth noting that databases were very different in the 1990s than they are today. Many of the problems that these standards tried to solve were aimed at row-oriented databases, which were prevalent at the time. Column-oriented databases arrived more than a decade later, and they have since come to dominate the analytics landscape. Unfortunately, without a column-oriented standard for transferring data, many of these databases had to retrofit a design that made them ODBC/JDBC-compatible. This allowed them to work with the countless database client tools in existence today but required a trade-off in performance in efficiency.

ADBC is the column-oriented specification that solves this problem. The pandas library, and many similar offerings in the space, are explicitly (or at least very close to) being column-oriented in their designs. When interacting with columnar databases like BigQuery, Redshift, or Snowflake, having a column-oriented driver to exchange information can lead to orders of magnitude better performance. Even if you aren't interacting with a column-oriented database, the ADBC driver is so finely optimized toward analytics with Apache Arrow that it *still* would be an upgrade over any ODBC/JDBC driver that SQLAlchemy would use.

For users wanting to know more about ADBC, I recommend viewing my talk from PyData NYC 2023, titled Faster SQL with pandas and Apache Arrow, on YouTube (`https://youtu.be/XhnfybpWOgA?si =RBrM7UUvpNFyct0L`).

Apache Parquet

As far as a generic storage format for a `pd.DataFrame` goes, Apache Parquet is the best option. Apache Parquet allows:

- Metadata storage – this allows the format to track data types, among other features
- Partitioning – not everything needs to be in one file
- Query support – Parquet files can be queried on disk, so you don't have to bring all data into memory
- Parallelization – reading data can be parallelized for higher throughput
- Compactness – data is compressed and stored in a highly efficient manner

Unless you are working with legacy systems, the Apache Parquet format should replace the use of CSV files in your workflows, from persisting data locally and sharing with other team members to exchanging data across systems.

How to do it

The API to read/write Apache Parquet is consistent with all other pandas APIs we have seen so far; for reading, there is pd.read_parquet, and for writing, there is a pd.DataFrame.to_parquet method.

Let's start with some sample data and an io.BytesIO object:

```
import io
buf = io.BytesIO()
df = pd.DataFrame([
    ["Paul", "McCartney", 1942],
    ["John", "Lennon", 1940],
    ["Richard", "Starkey", 1940],
    ["George", "Harrison", 1943],
], columns=["first", "last", "birth"])
df = df.convert_dtypes(dtype_backend="numpy_nullable")
df
```

```
     first    last       birth
0    Paul     McCartney  1942
1    John     Lennon     1940
2    Richard  Starkey    1940
3    George   Harrison   1943
```

This is how you would write to a file handle:

```
df.to_parquet(buf, index=False)
```

And here is how you would read from a file handle. Note that we are intentionally not providing dtype_backend="numpy_nullable":

```
buf.seek(0)
pd.read_parquet(buf)
```

```
     first    last       birth
0    Paul     McCartney  1942
1    John     Lennon     1940
2    Richard  Starkey    1940
3    George   Harrison   1943
```

Why don't we need the dtype_backend= argument with pd.read_parquet? Unlike a format like CSV, which only stores data, the Apache Parquet format stores both data and metadata. Within the metadata, Apache Parquet is able to keep track of the data types in use, so whatever data type you write should be exactly what you get back.

You can test this by changing the data type of the birth column to a different type:

```
df["birth"] = df["birth"].astype(pd.UInt16Dtype())
df.dtypes
```

```
first     string[python]
last      string[python]
birth              UInt16
dtype: object
```

Roundtripping this through the Apache Parquet format will give you back the same data type you started with:

```
buf = io.BytesIO()
df.to_parquet(buf, index=False)
buf.seek(0)
pd.read_parquet(buf).dtypes
```

```
first     string[python]
last      string[python]
birth              UInt16
dtype: object
```

Of course, if you want to be extra-defensive, there is no harm in using dtype_backend="numpy_nullable" here as well. We intentionally left it out at the start to showcase the power of the Apache Parquet format, but if you are receiving files from other sources and developers that don't use the type system we recommended in *Chapter 3, Data Types*, it may be helpful to make sure you work with the best types pandas has to offer:

```
suboptimal_df = pd.DataFrame([
    [0, "foo"],
    [1, "bar"],
    [2, "baz"],
], columns=["int_col", "str_col"])
buf = io.BytesIO()
suboptimal_df.to_parquet(buf, index=False)
buf.seek(0)
pd.read_parquet(buf, dtype_backend="numpy_nullable").dtypes
```

```
int_col             Int64
str_col    string[python]
dtype: object
```

Another great feature of the Apache Parquet format is that it supports *partitioning*, which loosens the requirement that all your data is located in a single file. By being able to split data across different directories and files, partitioning allows users an easy way to organize their content, while also making it easier for a program to optimize which files it may or may not have to read to solve an analytical query.

There are many ways to partition your data each with practical space/time trade-offs. For demonstration purposes, we are going to assume the use of *time-based partitioning*, whereby individual files are generated for different time periods. With that in mind, let's work with the following data layout, where we create different directories for each year and, within each year, create individual files for every quarter of sales:

```
Partitions
  2022/
    q1_sales.parquet
    q2_sales.parquet
  2023/
    q1_sales.parquet
    q2_sales.parquet
```

Each of the sample Apache Parquet files distributed with this book has already been created with the pandas extension types we recommended in *Chapter 3, Data Types* so the pd.read_parquet calls we make intentionally do not include the dtype_backend="numpy_nullable" argument. Within any file, you will see that we store information about the year, quarter, region, and overall sales that were counted:

```
pd.read_parquet(
    "data/partitions/2022/q1_sales.parquet",
)
```

	year	quarter	region	sales
0	2022	Q1	America	1
1	2022	Q1	Europe	2

If we wanted to see all of this data together, a brute-force approach would involve looping over each file and accumulating the results. However, with the Apache Parquet format, pandas can natively and effectively handle this. Instead of passing individual file names to pd.read_parquet, it simply passes the path to the directory:

```
pd.read_parquet("data/partitions/")
```

	year	quarter	region	sales
0	2022	Q1	America	1
1	2022	Q1	Europe	2
2	2022	Q2	America	4
3	2022	Q2	Europe	8
4	2023	Q1	America	16
5	2023	Q1	Europe	32
6	2023	Q2	America	64
7	2023	Q2	Europe	128

Because our sample data is so small, we have no problem reading all of the data into a `pd.DataFrame` first and then working with it from there. However, in production deployments, you may end up working with Apache Parquet files that measure in gigabytes or terabytes' worth of storage. Attempting to read all of that data into a `pd.DataFrame` may throw a `MemoryError`.

Fortunately, the Apache Parquet format gives you the capability to filter records on the fly as files are read. From pandas, you can enable this functionality with `pd.read_parquet` by passing a `filters=` argument. The argument should be a list, where each list element is a tuple, which itself contains three elements:

- Column Name
- Logical Operator
- Value

For example, if we only wanted to read in data where our `region` column is equal to the value `Europe`, we could write this as:

```
pd.read_parquet(
    "data/partitions/",
    filters=[("region", "==", "Europe")],
)
```

```
     year    quarter    region    sales
0    2022    Q1         Europe    2
1    2022    Q2         Europe    8
2    2023    Q1         Europe    32
3    2023    Q2         Europe    128
```

JSON

JavaScript Object Notation (JSON) is a common format used to transfer data over the internet. The JSON specification can be found at `https://www.json.org`. Despite the name, it does not require JavaScript to read or create.

The Python standard library ships with the `json` library, which can serialize and deserialize Python objects to/from JSON:

```
import json
beatles = {
    "first": ["Paul", "John", "Richard", "George",],
    "last": ["McCartney", "Lennon", "Starkey", "Harrison",],
    "birth": [1942, 1940, 1940, 1943],
}
serialized = json.dumps(beatles)
print(f"serialized values are: {serialized}")
deserialized = json.loads(serialized)
print(f"deserialized values are: {deserialized}")
```

```
serialized values are: {"first": ["Paul", "John", "Richard", "George"], "last":
["McCartney", "Lennon", "Starkey", "Harrison"], "birth": [1942, 1940, 1940,
1943]}
deserialized values are: {'first': ['Paul', 'John', 'Richard', 'George'],
'last': ['McCartney', 'Lennon', 'Starkey', 'Harrison'], 'birth': [1942, 1940,
1940, 1943]}
```

However, the standard library does not know how to deal with pandas objects, so pandas provides its own set of I/O functions specifically for JSON.

How to do it

In the simplest form, pd.read_json can be used to read JSON data:

```
import io
data = io.StringIO(serialized)
pd.read_json(data, dtype_backend="numpy_nullable")
```

```
     first    last       birth
0    Paul     McCartney  1942
1    John     Lennon     1940
2    Richard  Starkey    1940
3    George   Harrison   1943
```

And the pd.DataFrame.to_json method can be used for writing:

```
df = pd.DataFrame(beatles)
print(df.to_json())
```

```
{"first":{"0":"Paul","1":"John","2":"Richard","3":"George"},"last":{"0":"Mc-
Cartney","1":"Lennon","2":"Starkey","3":"Harrison"},"birth":{"0":1942,"1":1-
940,"2":1940,"3":1943}}
```

However, in practice, there are endless ways to represent tabular data in JSON. Some users may want to see each row of the pd.DataFrame represented as a JSON array, whereas other users may want to see each column shown as an array. Others may want to see the row index, column index, and data listed as separate JSON objects, whereas others may not care about seeing the row or column labels at all.

For these use cases and more, pandas allows you to pass an argument to orient=, whose value dictates the layout of the JSON to be read or written:

- columns (default): Produces JSON objects, where the key is a column label and the value is another object that maps the row label to a data point.
- records: Each row of the pd.DataFrame is represented as a JSON array, containing objects that map column names to a data point.
- split: Maps to {"columns": […], "index": […], "data": […]}. Columns/index values are arrays of labels, and data contains arrays of arrays.

- index: Similar to columns, except that the usage of row and column labels as keys is reversed.
- values: Maps the data of a pd.DataFrame to an array of arrays. Row/column labels are dropped.
- table: Adheres to the JSON Table Schema.

JSON is a *lossy* format for exchanging data, so each of the orients above is a trade-off between loss, verbosity, and end user requirements. orient="table" would be the least lossy and produce the largest payload, whereas orient="values" falls completely on the other end of that spectrum.

To highlight the differences in each of these orients, let's begin with a rather simple pd.DataFrame:

```python
df = pd.DataFrame(beatles, index=["row 0", "row 1", "row 2", "row 3"])
df = df.convert_dtypes(dtype_backend="numpy_nullable")
df
```

	first	last	birth
row 0	Paul	McCartney	1942
row 1	John	Lennon	1940
row 2	Richard	Starkey	1940
row 3	George	Harrison	1943

Passing orient="columns" will produce data using the pattern of {"column":{"row": value, "row": value, ...}, ...}:

```python
serialized = df.to_json(orient="columns")
print(f'Length of orient="columns": {len(serialized)}')
serialized[:100]
```

```
Length of orient="columns": 221
{"first":{"row 0":"Paul","row 1":"John","row 2":"Richard","row
3":"George"},"last":{"row 0":"McCartn
```

This is a rather verbose way of storing the data, as it will repeat the row index labels for every column. On the plus side, pandas can do a reasonably good job of reconstructing the proper pd.DataFrame from this orient:

```python
pd.read_json(
    io.StringIO(serialized),
    orient="columns",
    dtype_backend="numpy_nullable"
)
```

	first	last	birth
row 0	Paul	McCartney	1942
row 1	John	Lennon	1940
row 2	Richard	Starkey	1940
row 3	George	Harrison	1943

With `orient="records"`, you end up with each row of the `pd.DataFrame` being represented without its row index label, yielding a pattern of [{"col": value, "col": value, ...}, ...]:

```
serialized = df.to_json(orient="records")
print(f'Length of orient="records": {len(serialized)}')
serialized[:100]
```

```
Length of orient="records": 196
[{"first":"Paul","last":"McCartney","birth":1942},{"first":"John","last":"Len-
non","birth":1940},{"fi
```

While this representation is more compact than `orient="columns"`, it does not store any row labels, so on reconstruction, you will get back a `pd.DataFrame` with a newly generated `pd.RangeIndex`:

```
pd.read_json(
    io.StringIO(serialized),
    orient="orient",
    dtype_backend="numpy_nullable"
)
```

```
     first     last       birth
0    Paul      McCartney  1942
1    John      Lennon     1940
2    Richard   Starkey    1940
3    George    Harrison   1943
```

With `orient="split"`, the row index labels, column index labels, and data are all stored separately:

```
serialized = df.to_json(orient="split")
print(f'Length of orient="split": {len(serialized)}')
serialized[:100]
```

```
Length of orient="split": 190
{"columns":["first","last","birth"],"index":["row 0","row 1","row 2","row
3"],"data":[["Paul","McCar
```

This format uses a relatively lesser amount of characters than `orient="columns"`, and you can still recreate a `pd.DataFrame` reasonably well, since it mirrors how you would build a `pd.DataFrame` using the constructor (with arguments like `pd.DataFrame(data, index=index, columns=columns)`):

```
pd.read_json(
    io.StringIO(serialized),
    orient="split",
    dtype_backend="numpy_nullable",
)
```

```
row 0   Paul      McCartney  1942
```

```
row 1    John        Lennon      1940
row 2    Richard     Starkey     1940
row 3    George      Harrison    1943
```

While this is a good format when roundtripping a pd.DataFrame, the odds of coming across this JSON format "in the wild" are much lower as compared to other formats.

orient="index" is very similar to orient="columns", but it reverses the roles of the row and column labels:

```
serialized = df.to_json(orient="index")
print(f'Length of orient="index": {len(serialized)}')
serialized[:100]
```

```
Length of orient="index": 228
{"row 0":{"first":"Paul","last":"McCartney","birth":1942},"row 1":{"first":"-
John","last":"Lennon","b
```

Once again, you can recreate your pd.DataFrame reasonably well:

```
pd.read_json(
    io.StringIO(serialized),
    orient="index",
    dtype_backend="numpy_nullable",
)
```

```
         first    last        birth
row 0    Paul     McCartney   1942
row 1    John     Lennon      1940
row 2    Richard  Starkey     1940
row 3    George   Harrison    1943
```

Generally, orient="index" will take up more space than orient="columns", since most pd.DataFrame objects use column labels that are more verbose than index labels. I would only advise using this format in the possibly rare instances where your column labels are less verbose, or if you have strict formatting requirements imposed by another system.

For the most minimalistic representation, you can opt for orient="values". With this orient, neither row nor column labels are preserved:

```
serialized = df.to_json(orient="values")
print(f'Length of orient="values": {len(serialized)}')
serialized[:100]
```

```
Length of orient="values": 104
[["Paul","McCartney",1942],["John","Lennon",1940],["Richard","Star-
key",1940],["George","Harrison",19
```

Of course, since they are not represented in the JSON data, you will not maintain row/column labels when reading with `orient="values"`:

```
pd.read_json(
    io.StringIO(serialized),
    orient="values",
    dtype_backend="numpy_nullable",
)
```

	0	1	2
0	Paul	McCartney	1942
1	John	Lennon	1940
2	Richard	Starkey	1940
3	George	Harrison	1943

Finally, we have `orient="table"`. This will be the most verbose out of all of the outputs, but it is the only one backed by an actual standard, which is called the JSON Table Schema:

```
serialized = df.to_json(orient="table")
print(f'Length of orient="table": {len(serialized)}')
serialized[:100]
```

```
Length of orient="table": 524
{"schema":{"fields":[{"name":"index","type":"string"},{"name":"first","-
type":"any","extDtype":"strin
```

The Table Schema is more verbose because it stores metadata about the data being serialized, similar to what we saw with the Apache Parquet format (although with fewer features than Apache Parquet). With all of the other `orient=` arguments, pandas would have to infer the type of data as it is being read, but the JSON Table Format preserves that information for you. As such, you don't even need the `dtype_backend="numpy_nullable"` argument, assuming you used the pandas extension types to begin with:

```
df["birth"] = df["birth"].astype(pd.UInt16Dtype())
serialized = df.to_json(orient="table")
pd.read_json(
    io.StringIO(serialized),
    orient="table",
).dtypes
```

```
first     string[python]
last      string[python]
birth             UInt16
dtype: object
```

There's more...

When attempting to read JSON, you may find that none of the above formats can still sufficiently express what you are trying to accomplish. Fortunately, there is still `pd.json_normalize`, which can act as a workhorse function to convert your JSON data into a tabular format.

Imagine working with the following JSON data from a theoretical REST API with pagination:

```
data = {
    "records": [{
        "name": "human",
        "characteristics": {
            "num_leg": 2,
            "num_eyes": 2
        }
    }, {
        "name": "dog",
        "characteristics": {
            "num_leg": 4,
            "num_eyes": 2
        }
    }, {
        "name": "horseshoe crab",
        "characteristics": {
            "num_leg": 10,
            "num_eyes": 10
        }
    }],
    "type": "animal",
    "pagination": {
        "next": "23978sdlkusdf97234u2io",
        "has_more": 1
    }
}
```

While the "pagination" key is useful for navigating the API, it is of little reporting value to us and can trip up the JSON serializers. What we actually care about is the array associated with the "records" key. You can direct pd.json_normalize to look at this data exclusively, using the record_path= argument. Please note that pd.json_normalize is not a true I/O function, since it deals with Python objects and not file handles, so it has no dtype_backend= argument; instead, we will chain in a pd.DataFrame. convert_dtypes call to get the desired pandas extension types:

```
pd.json_normalize(
    data,
    record_path="records"
).convert_dtypes(dtype_backend="numpy_nullable")
```

	name	characteristics.num_leg	characteristics.num_eyes
0	human	2	2
1	dog	4	2
2	horseshoe crab	10	10

By providing the record_path= argument, we were able to ignore the undesired "pagination" key, but unfortunately, we now have the side effect of dropping the "type" key, which contained valuable metadata about each record. To preserve this information, you can use the meta= argument:

```
pd.json_normalize(
    data,
    record_path="records",
    meta="type"
).convert_dtypes(dtype_backend="numpy_nullable")
```

	name	characteristics.num_leg	characteristics.num_eyes	type
0	human	2	2	animal
1	dog	4	2	animal
2	horseshoe crab	10	10	animal

HTML

You can use pandas to read HTML tables from websites. This makes it easy to ingest tables such as those found on Wikipedia.

In this recipe, we will scrape tables from the Wikipedia entry for *The Beatles Discography* (https://en.wikipedia.org/wiki/The_Beatles_discography). In particular, we want to scrape the table in the image that was on Wikipedia in 2024:

List of studio albums, with selected chart positions and certification

Title	Album details[A]	Peak chart positions							Certifications
		UK [7] [8]	AUS [9]	CAN [10]	FRA [11]	GER [12]	NOR [13]	US [14] [15]	
Please Please Me	• Released: 22 March 1963 • Label: Parlophone	1	—	—	5	5	—	155	BPI: Platinum[16] ARIA: Gold[17] MC: Gold[18] RIAA: Platinum[19]
With the Beatles[B]	• Released: 22 November 1963 • Label: Parlophone (UK), Capitol (Canada), Odeon (France)	1	—	—	5	1	—	179	BPI: Gold[16] ARIA: Gold[17] BVMI: Gold[20] MC: Gold[18] RIAA: Gold[19]

Figure 4.4: Wikipedia page for The Beatles Discography

Before attempting to read HTML, users will need to install a third-party library. For the examples in this section, we will use lxml:

```
python -m pip install lxml
```

How to do it

pd.read_html allows you to read a table from a website:

```
url = "https://en.wikipedia.org/wiki/The_Beatles_discography"
dfs = pd.read_html(url, dtype_backend="numpy_nullable")
len(dfs)
```

Contrary to the other I/O methods we have seen so far, pd.read_html doesn't return a pd.DataFrame but, instead, returns a list of pd.DataFrame objects. Let's see what the first list element looks like:

```
dfs[0]
```

	The Beatles discography	The Beatles discography.1
0	The Beatles in 1965	The Beatles in 1965
1	Studio albums	12 (UK), 17 (US)
2	Live albums	5
3	Compilation albums	51
4	Video albums	22
5	Music videos	53
6	EPs	36
7	Singles	63
8	Mash-ups	2
9	Box sets	17

The preceding table is a summary of the count of studio albums, live albums, compilation albums, and so on. This is not the table we wanted. We could loop through each of the tables that pd.read_html created, or we could give it a hint to find a specific table.

One way of getting the table we want would be to leverage the attrs= argument of pd.read_html. This parameter accepts a dictionary mapping HTML attributes to values. Because an id attribute in HTML is supposed to be unique within a document, trying to find a table with attrs={"id": ...} is usually a safe approach. Let's see if we can get that to work here.

Use your web browser to inspect the HTML of the web page (if you are unsure how to do this, search online for terms like *Firefox inspector*, *Safari Web Inspector*, or *Google Chrome DevTools*; the terminology is unfortunately not standardized). Look for any id fields, unique strings, or attributes of the table element that help us identify the table we are after.

Here is a portion of the raw HTML:

```
<table class="wikipedia plainrowheaders" style="text-align:center;">
  <caption>List of studio albums, with selected chart positions and certifica-
tion
  </caption>
  <tbody>
    <tr>
      <th rowspan="2" scope="col" style="width:20em;">Title</th>
      <th rowspan="2" scope="col" style="width:20em;">Album details<sup
id="cite_ref-8" class="reference"><a href="#cite_note-8">[A]</a></sup></th>
      ...
    </tr>
  </tbody>
```

Unfortunately, the table we are looking for does *not* have an id attribute. We could try using either the class or style attributes we see in the HTML snippet above, but chances are those won't be unique.

Another parameter we can try is match=, which can be a string or a regular expression and matches against the table contents. In the <caption> tag of the above HTML, you will see the text "List of studio albums"; let's try that as an argument. To help readability, we are just going to look at each Album and its performance in the UK, AUS, and CAN:

```
url = "https://en.wikipedia.org/wiki/The_Beatles_discography"
dfs = pd.read_html(
    url,
    match=r"List of studio albums",
    dtype_backend="numpy_nullable",
)
print(f"Number of tables returned was: {len(dfs)}")
dfs[0].filter(regex=r"Title|UK|AUS|CAN").head()
```

	Title	Peak chart positions		
	Title	UK [8][9]	AUS [10]	CAN [11]
0	Please Please Me	1	–	–
1	With the Beatles[B]	1	–	–
2	A Hard Day's Night	1	1	–
3	Beatles for Sale	1	1	–
4	Help!	1	1	–

While we are able to now find the table, the column names are less than ideal. If you look closely at the Wikipedia table, you will notice that it partially creates a hierarchy between the Peak chart positions text and the name of countries below it, which pandas turns into a pd.MultiIndex. To make our table easier to read, we can pass header=1 to ignore the very first level of the generated pd.MultiIndex:

```
url = "https://en.wikipedia.org/wiki/The_Beatles_discography"
dfs = pd.read_html(
    url,
    match="List of studio albums",
    header=1,
    dtype_backend="numpy_nullable",
)
dfs[0].filter(regex=r"Title|UK|AUS|CAN").head()
```

	Title	UK [8][9]	AUS [10]	CAN [11]
0	Please Please Me	1	–	–
1	With the Beatles[B]	1	–	–
2	A Hard Day's Night	1	1	–

```
3      Beatles for Sale              1        1        —
4      Help!                         1        1        —
```

As we look closer at the data, we can see that Wikipedia uses – to represent missing values. If we pass this as an argument to the na_values= parameter of pd.read_html, we will see the =–= values converted to missing values:

```python
url = "https://en.wikipedia.org/wiki/The_Beatles_discography"
dfs = pd.read_html(
    url,
    match="List of studio albums",
    header=1,
    na_values=["–"],
    dtype_backend="numpy_nullable",
)
dfs[0].head()
```

```
        Title    Album details[A]        UK [8][9]       AUS [10]         CAN
[11]         FRA [12]        GER [13]        NOR [14]        US [15][16]
Certifications    Sales
0        Please Please Me         Released: 22 March 1963 Label: Parlophone
1        <NA>    <NA>    5        5        <NA>    155        BPI: Platinum[17] ARIA:
Gold[18] MC: Gold[19] …    <NA>
1    With the Beatles[B]        Released: 22 November 1963 Label: Parlophone (…  1
<NA>    <NA>    5        1        <NA>    179        BPI: Gold[17] ARIA: Gold[18]
BVMI: Gold[21] MC… <NA>
2    A Hard Day's Night        Released: 10 July 1964 Label: Parlophone 1        1
<NA>    <NA>    1        <NA>    <NA>    BPI: Platinum[17] ARIA: Gold[18]
<NA>
3    Beatles for Sale         Released: 4 December 1964 Label: Parlophone        1
1        <NA>    <NA>    1        <NA>    <NA>    BPI: Gold[17] ARIA: Gold[18]
MC: Gold[19] RIAA… UK: 750,000[22]
4        Help!    Released: 6 August 1965 Label: Parlophone        1        1
<NA>    5        1        <NA>    <NA>    BPI: Platinum[17] ARIA: Gold[18]
<NA>
```

Pickle

The pickle format is Python's built-in serialization format. Pickle files typically end with a .pkl extension.

Unlike other formats encountered so far, the pickle format should not be used to transfer data across machines. The main use case is for saving pandas objects *that themselves contain Python objects* to your own machine, returning to them at a later point in time. If you are unsure if you should be using this format or not, I would advise trying the Apache Parquet format first, which covers a wider array of use cases.

Do not load pickle files from untrusted sources. I would generally only advise using pickle for your own analyses; do not share data or expect to receive data from others in the pickle format.

How to do it

To highlight that the pickle format should really only be used when your pandas objects contain Python objects, let's imagine we decided to store our Beatles data as a pd.Series of namedtuple types. It is a fair question as to *why* you would do this in the first place, as it would be better represented as a pd.DataFrame... but questions aside, it is valid to do so:

```python
from collections import namedtuple

Member = namedtuple("Member", ["first", "last", "birth"])
ser = pd.Series([
    Member("Paul", "McCartney", 1942),
    Member("John", "Lennon", 1940),
    Member("Richard", "Starkey", 1940),
    Member("George", "Harrison", 1943),
])
ser
```

```
0       (Paul, McCartney, 1942)
1           (John, Lennon, 1940)
2       (Richard, Starkey, 1940)
3       (George, Harrison, 1943)
dtype: object
```

None of the other I/O methods we have discussed in this chapter would be able to faithfully represent a namedtuple, which is purely a Python construct. pd.Series.to_pickle, however, has no problem writing this out:

```python
import io
buf = io.BytesIO()
ser.to_pickle(buf)
```

When you call pd.read_pickle, you will get the exact representation you started with returned:

```python
buf.seek(0)
ser = pd.read_pickle(buf)
ser
```

```
0       (Paul, McCartney, 1942)
1           (John, Lennon, 1940)
2       (Richard, Starkey, 1940)
3       (George, Harrison, 1943)
dtype: object
```

You can further validate this by inspecting an individual element:

```
ser.iloc[0]
```

```
Member(first='Paul', last='McCartney', birth=1942)
```

Once again, it is worth stressing that the Apache Parquet format should be preferred to pickle, only using this as a last resort when Python-specific objects within your pd.Series or pd.DataFrame need to be roundtripped. Be sure to **never load pickle files from untrusted sources**; unless you created the pickle file yourself, I would highly advise against trying to process it.

Third-party I/O libraries

While pandas covers an impressive amount of formats it cannot hope to cover *every* important format out there. Third-party libraries exist to cover that gap.

Here are a few you may be interested in – the details of how they work are outside the scope of this book, but they all generally follow the pattern of having read functions that return pd.DataFrame objects and write methods that accept a pd.DataFrame argument:

- pandas-gbq allows you to exchange data with Google BigQuery
- AWS SDK for pandas works with Redshift and the AWS ecosystem at large
- Snowflake Connector for Python helps exchange with Snowflake databases
- pantab lets you move pd.DataFrame objects in and out of Tableau's Hyper database format (note: I am also the author of pantab)

Unlock this book's exclusive benefits now

This book comes with additional benefits designed to elevate your learning experience.

Note: Have your purchase invoice ready before you begin. https://www.packtpub.com/unlock/9781836205876

5

Algorithms and How to Apply Them

In this book, we have already looked at a variety of ways to create pandas data structures and select/ assign data within them and have subsequently seen how to store those structures in common formats. These features alone can make pandas a powerful tool in the realm of data exchange, but we are still just scratching the surface of what pandas can offer.

A core component of data analysis and computing in general is the application of *algorithms*, which describe a sequence of steps the computer should take to process data. In their simplistic form, common data algorithms build upon basic arithmetic (for example, "sum this column"), but scale out to any sequence of steps that you may need for your custom calculations.

As you will see in this chapter, pandas provides many common data algorithms out of the box, but also gives you a robust framework through which you can compose and apply your own algorithms. The algorithms pandas provides out of the box would be faster than anything you can write by hand in Python, and as you progress in your data journey, you will usually find that clever use of these algorithms can cover a vast amount of data processing needs.

We are going to cover the following recipes in this chapter:

- Basic `pd.Series` arithmetic
- Basic `pd.DataFrame` arithmetic
- Aggregations
- Transformations
- Map
- Apply
- Summary statistics
- Binning algorithms
- One-hot encoding with `pd.get_dummies`

- Chaining with `.pipe`
- Selecting the lowest-budget movies from the top 100
- Calculating a trailing stop order price
- Finding the baseball players best at...
- Understanding which position scores the most per team

Basic pd.Series arithmetic

The easiest place to start when exploring pandas algorithms is with a `pd.Series`, given it is also the most basic structure provided by the pandas library. Basic arithmetic will cover the operations of addition, subtraction, multiplication, and division, and, as you will see in this section, pandas offers two ways to perform these. The first approach allows pandas to work with the +, -, *, and / operators built into the Python language, which is an intuitive way for new users coming to the library to pick up the tool. However, to cover features specific to data analysis not covered by the Python language, and to support the *Chaining with .pipe* approach that we will cover later in this chapter, pandas also offers `pd.Series.add`, `pd.Series.sub`, `pd.Series.mul`, and `pd.Series.div`, respectively.

The pandas library goes to great lengths to keep its API consistent across all data structures, so you will see that the knowledge from this section can be easily transferred over to the `pd.DataFrame` structure, with the only difference being that a `pd.Series` is one-dimensional while a `pd.DataFrame` is two-dimensional.

How to do it

Let's create a simple `pd.Series` from a Python `range` expression:

```
ser = pd.Series(range(3), dtype=pd.Int64Dtype())
ser
```

```
0    0
1    1
2    2
dtype: Int64
```

To establish terminology, let's briefly consider an expression like a + b. In such an expression, we are using a *binary operator* (+). The term *binary* refers to the fact that you need to add two things together for this expression to make sense, that is, it wouldn't make sense to just have an expression like a +. Those two "things" are technically considered *operands*; so, with a + b, we have a left operand of a and a right operand of b.

With one of the operands being a `pd.Series`, the most basic algorithmic expression in pandas would encompass the other operand being a *scalar*, that is to say, just a single value. When that occurs, the scalar value is *broadcast* to each element of the `pd.Series` to apply the algorithm.

For example, if we wanted to add the number 42 to each and every element of our `pd.Series`, we could simply express that as:

```
ser + 42
```

```
0    42
1    43
2    44
dtype: Int64
```

The pandas library is able to take the addition expression and apply it to our pd.Series in a *vectorized* manner (i.e., the number 42 gets applied to all values at once without requiring users to resort to a for loop in Python).

Subtraction may be expressed naturally using the - operator:

```
ser - 42
```

```
0    -42
1    -41
2    -40
dtype: Int64
```

Similarly, multiplication may be expressed with the * operator:

```
ser * 2
```

```
0    0
1    2
2    4
dtype: Int64
```

By now, you can probably surmise that division is expressed with the / operator:

```
ser / 2
```

```
0    0.0
1    0.5
2    1.0
dtype: Float64
```

It is also perfectly valid for the two operands to be a pd.Series:

```
ser2 = pd.Series(range(10, 13), dtype=pd.Int64Dtype())
ser + ser2
```

```
0    10
1    12
2    14
dtype: Int64
```

As mentioned in the introduction of this section, while the built-in Python operators are commonly used and viable in most cases, pandas still offers dedicated methods for pd.Series.add, pd.Series. sub, pd.Series.mul, and pd.Series.div:

```
ser1 = pd.Series([1., 2., 3.], dtype=pd.Float64Dtype())
ser2 = pd.Series([4., pd.NA, 6.], dtype=pd.Float64Dtype())
ser1.add(ser2)
```

```
0    5.0
1    <NA>
2    9.0
dtype: Float64
```

The advantage of pd.Series.add over the built-in operator is that it accepts an optional fill_value= argument to handle missing data:

```
ser1.add(ser2, fill_value=0.)
```

```
0    5.0
1    2.0
2    9.0
dtype: Float64
```

Later in this chapter, you will also be introduced to chaining with .pipe, which chains most naturally with the pandas methods and not with the built-in Python operators.

There's more...

When both operands in your expression are pd.Series objects together, it is important to note that pandas will align on the row labels. This alignment behavior is considered a feature, but can also be surprising to newcomers.

To see why this matters, let's start with two pd.Series objects that have an identical row index. When we try to add these together, we get a rather unsurprising result:

```
ser1 = pd.Series(range(3), dtype=pd.Int64Dtype())
ser2 = pd.Series(range(3), dtype=pd.Int64Dtype())
ser1 + ser2
```

```
0    0
1    2
2    4
dtype: Int64
```

But what happens when the row index values are not identical? A simple case may involve adding two pd.Series objects together, where one pd.Series uses a row index that is a subset of the other. You can see this with ser3 in the following code, which only has 2 values and uses the default pd.RangeIndex with values of [0, 1]. When added together with ser1, we still get a 3-element pd.Series in return, but values are only added where the row index labels can be aligned from both pd.Series objects:

```
ser3 = pd.Series([2, 4], dtype=pd.Int64Dtype())
ser1 + ser3
```

```
0       2
1       5
2    <NA>
dtype: Int64
```

Now let's take a look at what happens when two pd.Series objects of the same length get added together, but the row index values are different:

```
ser4 = pd.Series([2, 4, 8], index=[1, 2, 3], dtype=pd.Int64Dtype())
ser1 + ser4
```

```
0    <NA>
1       3
2       6
3    <NA>
dtype: Int64
```

For an even more extreme case, let's consider the situation where one pd.Series has row index values that are non-unique:

```
ser5 = pd.Series([2, 4, 8], index=[0, 1, 1], dtype=pd.Int64Dtype())
ser1 + ser5
```

```
0       2
1       5
1       9
2    <NA>
dtype: Int64
```

If you have a background in SQL, the behavior of pandas here is akin to a FULL OUTER JOIN in a database. Every label from each row index gets included in the output, with pandas matching up the labels that can be seen in both pd.Series objects. This can be directly replicated in a database like PostgreSQL:

```
WITH ser1 AS (
  SELECT * FROM (
    VALUES
      (0, 0),
      (1, 1),
      (2, 2)
  ) AS t(index, val1)
),

ser5 AS (
  SELECT * FROM (
```

```
    VALUES
        (0, 2),
        (1, 4),
        (1, 8)
    ) AS t(index, val2)
)


SELECT * FROM ser1 FULL OUTER JOIN ser5 USING(index);
```

If you were to run this snippet directly in PostgreSQL, you would get back the following result:

```
index | val1 | val2
------+------+------
    0 |    0 |    2
    1 |    1 |    8
    1 |    1 |    4
    2 |    2 |
(4 rows)
```

Ignoring the ordering difference, you can see that the database gives us back all of the unique index values from the combinations of [0, 1, 2] and [0, 1, 1], alongside any associated val1 and val2 values. Even though ser1 only had one index value of 1, that same value appeared twice in the index column in ser5. The FULL OUTER JOIN therefore shows both val2 values from ser5 (4 and 8), while duplicating the val1 value originating from ser1 (1).

If you were to subsequently add val1 and val2 together in the database, you would get back a result that matches the output of ser1 + ser5, sparing the fact that the database may choose a different order for its output:

```
WITH ser1 AS (
  SELECT * FROM (
    VALUES
        (0, 0),
        (1, 1),
        (2, 2)
    ) AS t(index, val1)
),


ser5 AS (
  SELECT * FROM (
    VALUES
        (0, 2),
        (1, 4),
        (1, 8)
    ) AS t(index, val2)
```

```
)

SELECT index, val1 + val2 AS value FROM ser1 FULL OUTER JOIN ser5 USING(index);

index | value
------+-------
    0 |     2
    1 |     9
    1 |     5
    2 |
(4 rows)
```

Basic pd.DataFrame arithmetic

Having now covered basic pd.Series arithmetic, you will find that the corresponding pd.DataFrame arithmetic operations are practically identical, with the lone exception being that our algorithms now work in two dimensions instead of just one. In doing so, the pandas API makes it easy to interpret data regardless of its shape, and without requiring users to write loops to interact with data. This helps significantly reduce developer effort and helps you write faster code – a win-win for developers.

How it works

Let's create a small 3x3 pd.DataFrame using random numbers:

```
np.random.seed(42)
df = pd.DataFrame(
    np.random.randn(3, 3),
    columns=["col1", "col2", "col3"],
    index=["row1", "row2", "row3"],
).convert_dtypes(dtype_backend="numpy_nullable")
df
```

	col1	col2	col3
row1	0.496714	-0.138264	0.647689
row2	1.52303	-0.234153	-0.234137
row3	1.579213	0.767435	-0.469474

Much like a pd.Series, a pd.DataFrame also supports built-in binary operators with a scalar argument. Here is a simplistic addition operation:

```
df + 1
```

	col1	col2	col3
row1	1.496714	0.861736	1.647689
row2	2.52303	0.765847	0.765863
row3	2.579213	1.767435	0.530526

And here is a simplistic multiplication operation:

```
df * 2
```

	col1	col2	col3
row1	0.993428	-0.276529	1.295377
row2	3.04606	-0.468307	-0.468274
row3	3.158426	1.534869	-0.938949

You can also perform arithmetic with a pd.Series. By default, each row label in the pd.Series is searched for and aligned against the columns of the pd.DataFrame. To illustrate, let's create a small pd.Series whose index labels match the column labels of df:

```
ser = pd.Series(
    [20, 10, 0],
    index=["col1", "col2", "col3"],
    dtype=pd.Int64Dtype(),
)
ser
```

```
col1    20
col2    10
col3     0
dtype: Int64
```

If you were to try and add this to our pd.DataFrame, it would take the value of col1 in the pd.Series and add it to every element in the col1 column of the pd.DataFrame, repeating for each index entry:

```
df + ser
```

	col1	col2	col3
row1	20.496714	9.861736	0.647689
row2	21.52303	9.765847	-0.234137
row3	21.579213	10.767435	-0.469474

In cases where the row labels of the pd.Series do not match the column labels of the pd.DataFrame, you may end up with missing data:

```
ser = pd.Series(
    [20, 10, 0, 42],
    index=["col1", "col2", "col3", "new_column"],
    dtype=pd.Int64Dtype(),
)
ser + df
```

	col1	col2	col3	new_column
row1	20.496714	9.861736	0.647689	NaN
row2	21.52303	9.765847	-0.234137	NaN
row3	21.579213	10.767435	-0.469474	NaN

If you would like to control how pd.Series and pd.DataFrame align, you can use the axis= parameter of methods like pd.DataFrame.add, pd.DataFrame.sub, pd.DataFrame.mul, and pd.DataFrame.div.

Let's see this in action by creating a new pd.Series using row labels that align better with the row labels of our pd.DataFrame:

```
ser = pd.Series(
    [20, 10, 0, 42],
    index=["row1", "row2", "row3", "row4"],
    dtype=pd.Int64Dtype(),
)
ser
```

```
row1    20
row2    10
row3     0
row4    42
dtype: Int64
```

Specifying df.add(ser, axis=0) will match up the row labels from both the pd.Series and pd.DataFrame:

```
df.add(ser, axis=0)
```

	col1	col2	col3
row1	20.496714	19.861736	20.647689
row2	11.52303	9.765847	9.765863
row3	1.579213	0.767435	-0.469474
row4	<NA>	<NA>	<NA>

You can also use two pd.DataFrame arguments as the operands of addition, subtraction, multiplication, and division. Here is how to multiply two pd.DataFrame objects together:

```
df * df
```

	col1	col2	col3
row1	0.246725	0.019117	0.4195
row2	2.31962	0.054828	0.05482
row3	2.493913	0.588956	0.220406

Of course, when doing this, you still need to be aware of the index alignment rules – items are always aligned by label and not by position!

Let's create a new 3x3 pd.DataFrame with different row and column labels to show this:

```
np.random.seed(42)
df2 = pd.DataFrame(np.random.randn(3, 3))
df2 = df2.convert_dtypes(dtype_backend="numpy_nullable")
df2
```

	0	1	2
0	0.496714	-0.138264	0.647689
1	1.52303	-0.234153	-0.234137
2	1.579213	0.767435	-0.469474

Attempting to add this to our previous pd.DataFrame will generate a row index with labels ["row1", "row2", "row3", 0, 1, 2] and a column index with labels ["col1", "col2", "col3", 0, 1, 2]. Because no labels could be aligned, everything comes back as a missing value:

```
df + df2
```

	col1	col2	col3	0	1	2
row1	<NA>	<NA>	<NA>	<NA>	<NA>	<NA>
row2	<NA>	<NA>	<NA>	<NA>	<NA>	<NA>
row3	<NA>	<NA>	<NA>	<NA>	<NA>	<NA>
0	<NA>	<NA>	<NA>	<NA>	<NA>	<NA>
1	<NA>	<NA>	<NA>	<NA>	<NA>	<NA>
2	<NA>	<NA>	<NA>	<NA>	<NA>	<NA>

Aggregations

Aggregations (also referred to as *reductions*) help you to reduce multiple values from a series of values down to a single value. Even if the technical term is new to you, you have no doubt encountered many aggregations in your data journey. Things like the *count* of records, the *sum* or sales, or the *average* price are all very common aggregations.

In this recipe, we will explore many of the aggregations built into pandas, while also forming an understanding of how these aggregations are applied. Most analysis you will do throughout your data journey involves taking large datasets and aggregating the values therein into results that your audience can consume. Executives at most companies are not interested in receiving a data dump of transactions, they just want to know the sum, min, max, mean, and so on of values within those transactions. As such, effective use and application of aggregations is a key component to converting your complex data transformation pipelines into simple outputs that others can use and act upon.

How to do it

Many basic aggregations are implemented as methods directly on the pd.Series object, which makes it trivial to calculate commonly desired outputs like the count, sum, max, and so on.

To kick off this recipe, let's once again start with a pd.Series containing random numbers:

```
np.random.seed(42)
ser = pd.Series(np.random.rand(10_000), dtype=pd.Float64Dtype())
```

The pandas library provides methods for many commonly used aggregations, like `pd.Series.count`, `pd.Series.mean`, `pd.Series.std`, `pd.Series.min`, `pd.Series.max`, and `pd.Series.sum`:

```
print(f"Count is: {ser.count()}")
print(f"Mean value is: {ser.mean()}")
print(f"Standard deviation is: {ser.std()}")
print(f"Minimum value is: {ser.min()}")
print(f"Maximum value is: {ser.max()}")
print(f"Summation is: {ser.sum()}")
```

```
Count is: 10000
Mean value is: 0.49415955768429964
Standard deviation is: 0.2876301265269928
Minimum value is: 1.1634755366141114e-05
Maximum value is: 0.9997176732861306
Summation is: 4941.595576842997
```

Instead of calling those methods directly, a more generic way to invoke these aggregations would be to use `pd.Series.agg`, providing the name of the aggregation you would like to perform as a string:

```
print(f"Count is: {ser.agg('count')}")
print(f"Mean value is: {ser.agg('mean')}")
print(f"Standard deviation is: {ser.agg('std')}")
print(f"Minimum value is: {ser.agg('min')}")
print(f"Maximum value is: {ser.agg('max')}")
print(f"Summation is: {ser.agg('sum')}")
```

```
Count is: 10000
Mean value is: 0.49415955768429964
Standard deviation is: 0.2876301265269928
Minimum value is: 1.1634755366141114e-05
Maximum value is: 0.9997176732861306
Summation is: 4941.595576842997
```

An advantage using `pd.Series.agg` is that it can perform multiple aggregations for you. For example, if you wanted to calculate the minimum and maximum of a field in one step, you could do this by providing a list to `pd.Series.agg`:

```
ser.agg(["min", "max"])
```

```
min     0.000012
max     0.999718
dtype: float64
```

Aggregating a pd.Series is straightforward because there is only one dimension to be aggregated. With a pd.DataFrame, there are two possible dimensions to aggregate along, so you have a few more considerations as an end user of the library.

To walk through this, let's go ahead and create a pd.DataFrame with random numbers:

```
np.random.seed(42)
df = pd.DataFrame(
    np.random.randn(10_000, 6),
    columns=list("abcdef"),
).convert_dtypes(dtype_backend="numpy_nullable")
df
```

```
          a          b          c          d          e          f
0      0.496714  -0.138264   0.647689   1.523030  -0.234153  -0.234137
1      1.579213   0.767435  -0.469474   0.542560  -0.463418  -0.465730
2      0.241962  -1.913280  -1.724918  -0.562288  -1.012831   0.314247
3     -0.908024  -1.412304   1.465649  -0.225776   0.067528  -1.424748
4     -0.544383   0.110923  -1.150994   0.375698  -0.600639  -0.291694
...        ...        ...        ...        ...        ...        ...
9995   1.951254   0.324704   1.937021  -0.125083   0.589664   0.869128
9996   0.624062  -0.317340  -1.636983   2.390878  -0.597118   2.670553
9997  -0.470192   1.511932   0.718306   0.764051  -0.495094  -0.273401
9998  -0.259206   0.274769  -0.084735  -0.406717  -0.815527  -0.716988
9999   0.533743  -0.701856  -1.099044   0.141010  -2.181973  -0.006398
10000 rows × 6 columns
```

By default, invoking an aggregation using a built-in method like pd.DataFrame.sum will apply *along the columns*, meaning each column is individually aggregated. After that, pandas will display the result of each column's aggregation as an entry in a pd.Series:

```
df.sum()
```

```
a     -21.365908
b      -7.963987
c     152.032992
d    -180.727498
e      29.399311
f      25.042078
dtype: Float64
```

If you would like to aggregate data in each row, you can specify the axis=1 argument, with the caveat being that pandas is way more optimized for axis=0 operations, so this has a chance of being *significantly slower* than aggregating columns. Even still, it is a rather unique feature of pandas that can be useful when performance is not the main concern:

```
df.sum(axis=1)
```

```
0          2.060878
1          1.490586
2         -4.657107
3         -2.437675
4         -2.101088
             ...
9995       5.54669
9996       3.134053
9997       1.755601
9998      -2.008404
9999      -3.314518
Length: 10000, dtype: Float64
```

Much like a pd.Series, a pd.DataFrame has a .agg method, which can be used to apply multiple aggregations at once:

```
df.agg(["min", "max"])
```

	a	b	c	d	e	f
min	-4.295391	-3.436062	-3.922400	-4.465604	-3.836656	-4.157734
max	3.602415	3.745379	3.727833	4.479084	3.691625	3.942331

There's more...

In the examples covered in the *How to do it* section, we passed functions as strings like min and max to .agg. This is great for simple functions, but for more complex cases, you can also pass in callable arguments. Each callable should accept a single argument pd.Series and reduce down to a scalar:

```
def mean_and_add_42(ser: pd.Series):
    return ser.mean() + 42


def mean_and_sub_42(ser: pd.Series):
    return ser.mean() - 42


np.random.seed(42)
ser = pd.Series(np.random.rand(10_000), dtype=pd.Float64Dtype())
ser.agg([mean_and_add_42, mean_and_sub_42])
```

```
mean_and_add_42     42.49416
mean_and_sub_42    -41.50584
dtype: float64
```

Transformations

Contrary to *aggregations*, transformations do not reduce an array of values to a single value but, rather, maintain the shape of the calling object. This particular recipe may seem rather mundane coming from the previous section on aggregations, but transformations and aggregations will end up being very complementary tools to calculate things like the "% total of group" later in the cookbook.

How to do it

Let's create a small pd.Series:

```
ser = pd.Series([-1, 0, 1], dtype=pd.Int64Dtype())
```

Much like we saw with pd.Series.agg before, pd.Series.transform can accept a list of functions to apply. However, whereas pd.Series.agg expected these functions to return a single value, pd.Series.transform expects these functions to return a pd.Series with the same index and shape:

```
def adds_one(ser: pd.Series) -> pd.Series:
    return ser + 1

ser.transform(["abs", adds_one])
```

```
     abs     adds_one
0    1       0
1    0       1
2    1       2
```

Much like pd.DataFrame.agg would *aggregate* each column by default, pd.DataFrame.transform will *transform* each column by default. Let's create a small pd.DataFrame to see this in action:

```
df = pd.DataFrame(
    np.arange(-5, 4, 1).reshape(3, -1)
).convert_dtypes(dtype_backend="numpy_nullable")
df
```

```
     0    1    2
0    -5   -4   -3
1    -2   -1   0
2    1    2    3
```

Sparing implementation details, calling something like df.transform("abs") will apply the absolute value function to each column individually before piecing back together the result as a pd.DataFrame:

```
df.transform("abs")
```

```
     0    1    2
0    5    4    3
```

```
1     2     1     0
2     1     2     3
```

If you were to pass multiple transformation functions to pd.DataFrame.transform, you will end up with a pd.MultiIndex:

```
def add_42(ser: pd.Series):
    return ser + 42


df.transform(["abs", add_42])
```

| | 0 | | 1 | | 2 | |
	abs	add_42	abs	add_42	abs	add_42
0	5	37	4	38	3	39
1	2	40	1	41	0	42
2	1	43	2	44	3	45

There's more...

As mentioned in the introduction to this recipe, transformations and aggregations can work naturally together alongside the GroupBy concept, which will be covered in *Chapter 8*, *Group By*. In particular, our *Group by basics* recipe will be helpful to compare/contrast aggregations to transformations and will highlight how transformations can be used to expressively and succinctly calculate "percent of group" calculations.

Map

The .agg and .transform methods we have seen so far apply to an entire *sequence* of values at once. Generally, in pandas, this is a good thing; it allows pandas to perform *vectorized* operations that are fast and computationally efficient.

Still, sometimes, you as an end user may decide that you want to trade performance for customization or finer-grained control. This is where the .map methods can come into the picture; .map helps you apply functions individually to each element of your pandas object.

How to do it

Let's assume we have a pd.Series of data that mixes together both numbers and lists of numbers:

```
ser = pd.Series([123.45, [100, 113], 142.0, [110, 113, 119]])
ser
```

```
0             123.45
1         [100, 113]
2              142.0
3    [110, 113, 119]
dtype: object
```

.agg or .transform are not suitable here because we do not have a uniform data type – we really have to inspect each element to make a decision on how to handle it.

For our analysis, let's assume that when we encounter a number, we are happy to return the value as is. If we encounter a list of values, we want to average out all of the values within that list and return that. A function implementing this feature would look as follows:

```
def custom_average(value):
    if isinstance(value, list):
        return sum(value) / len(value)

    return value
```

We can then apply this to each element of our pd.Series using pd.Series.map:

```
ser.map(custom_average)
```

```
0     123.45
1     106.50
2     142.00
3     114.00
dtype: float64
```

If we had a pd.DataFrame containing this type of data, pd.DataFrame.map would be able to apply this function just as well:

```
df = pd.DataFrame([
    [2., [1, 2], 3.],
    [[4, 5], 5, 7.],
    [1, 4, [1, 1, 5.5]],
])
df
```

```
          0        1              2
0        2.0    [1, 2]           3.0
1      [4, 5]      5             7.0
2        1        4        [1, 1, 5.5]
```

```
df.map(custom_average)
```

```
        0      1      2
0      2.0    1.5    3.0
1      4.5    5.0    7.0
2      1.0    4.0    2.5
```

There's more...

In the above example, instead of using pd.Series.map, you could have also used pd.Series.transform:

```
ser.transform(custom_average)
```

```
0    123.45
1    106.50
2    142.00
3    114.00
dtype: float64
```

However, you would *not* get the same results with pd.DataFrame.transform:

```
df.transform(custom_average)
```

```
        0       1            2
0     2.0  [1, 2]          3.0
1  [4, 5]       5          7.0
2       1       4  [1, 1, 5.5]
```

Why is this? Remember that .map explicitly applies a function to each element, regardless of if you are working with a pd.Series or pd.DataFrame. pd.Series.transform is also happy to apply a function to each element that it contains, but pd.DataFrame.transform essentially loops over each column and passes that column as an argument to the callable arguments.

Because our function is implemented as:

```
def custom_average(value):
    if isinstance(value, list):
        return sum(value) / len(value)

    return value
```

the isinstance(value, list) check fails when passed a pd.Series and you end up just returning the pd.Series itself. If we tweak our function slightly:

```
def custom_average(value):
    if isinstance(value, (pd.Series, pd.DataFrame)):
        raise TypeError("Received a pandas object - expected a single value!")
    if isinstance(value, list):
        return sum(value) / len(value)

    return value
```

then the behavior of pd.DataFrame.transform becomes more clear:

```
df.transform(custom_average)
```

```
TypeError: Received a pandas object - expected a single value!
```

While there may be conceptual overlap, generally, in your code, you should think of .map as working element-wise, whereas .agg and .transform will try as best as they can to work with larger sequences of data at once.

Apply

Apply is a commonly used method, to the point that I would argue it is *overused*. The .agg, .transform, and .map methods seen so far have relatively clear semantics (.agg reduces, .transform maintains shape, .map applies functions element-wise), but when you reach for .apply, you can mirror any of these. That flexibility may seem nice at first, but because .apply leaves it up to pandas to *do the right thing*, you are typically better off picking the most explicit methods to avoid surprises.

Even still, you will see a lot of code out in the wild (especially from users who did not read this book); so, understanding what it does and what its limitations are can be invaluable.

How to do it

Calling pd.Series.apply will make .apply act like .map (i.e., the function gets applied to each individual element of the pd.Series).

Let's take a look at a rather contrived function that prints out each element:

```
def debug_apply(value):
    print(f"Apply was called with value:\n{value}")
```

Funneling this through .apply:

```
ser = pd.Series(range(3), dtype=pd.Int64Dtype())
ser.apply(debug_apply)
```

```
Apply was called with value:
0
Apply was called with value:
1
Apply was called with value:
2
0    None
1    None
2    None
dtype: object
```

gives exactly the same behavior as pd.Series.map:

```
ser.map(debug_apply)
```

```
Apply was called with value:
0
Apply was called with value:
1
Apply was called with value:
2
0    None
1    None
2    None
dtype: object
```

pd.Series.apply works like a Python loop, calling the function for each element. Because our function returns nothing, our resulting pd.Series is a like-indexed array of None values.

Whereas pd.Series.apply works element-wise, pd.DataFrame.apply works across each column as a pd.Series. Let's see this in action with a pd.DataFrame of shape (3, 2):

```
df = pd.DataFrame(
    np.arange(6).reshape(3, -1),
    columns=list("ab"),
).convert_dtypes(dtype_backend="numpy_nullable")
df
```

	a	b
0	0	1
1	2	3
2	4	5

```
df.apply(debug_apply)
```

```
Apply was called with value:
0    0
1    2
2    4
Name: a, dtype: Int64
Apply was called with value:
0    1
1    3
2    5
Name: b, dtype: Int64
a    None
b    None
dtype: object
```

As you can see in the above output, the function was only called twice given the two columns of data, but it was applied three times with the pd.Series that had three rows.

Aside from how many times pd.DataFrame.apply actually applies the function, the shape of the return value can vary between mirroring .agg and .transform functionality. Our preceding example is closer to a .agg because it returns a single None value, but if we returned the element we printed, we would get behavior more like a .transform:

```
def debug_apply_and_return(value):
    print(value)
    return value
df.apply(debug_apply_and_return)
```

```
0    0
1    2
2    4
Name: a, dtype: Int64
0    1
1    3
2    5
Name: b, dtype: Int64
   a  b
0  0  1
1  2  3
2  4  5
```

If you find this confusing, you are not alone. Trusting pandas to *do the right thing* with .apply can be a risky proposition; I strongly advise users exhaust all options with .agg, .transform, or .map before reaching for .apply.

Summary statistics

Summary statistics provide a quick way to understand the basic properties and distribution of the data. In this section, we introduce two powerful pandas methods: pd.Series.value_counts and pd.Series.describe, which can serve as useful starting points for exploration.

How to do it

The pd.Series.value_counts method attaches frequency counts to each distinct data point, making it easy to see how often each value occurs. This is particularly useful for discrete data:

```
ser = pd.Series(["a", "b", "c", "a", "c", "a"], dtype=pd.StringDtype())
ser.value_counts()
```

```
a    3
c    2
b    1
Name: count, dtype: Int64
```

For continuous data, `pd.Series.describe` is a heap of calculations packaged together into one method call. Through invocation of this particular method, you can easily see the count, mean, minimum, and maximum, alongside a high-level distribution of your data:

```
ser = pd.Series([0, 42, 84], dtype=pd.Int64Dtype())
ser.describe()
```

```
count     3.0
mean     42.0
std      42.0
min       0.0
25%      21.0
50%      42.0
75%      63.0
max      84.0
dtype: Float64
```

By default, we will see our distribution summarized through the 25%, 50%, 75%, and max (or 100%) quartiles. If your data analysis was focused on a more particular part of the distribution, you could control what this method presents back by providing a `percentiles=` argument:

```
ser.describe(percentiles=[.10, .44, .67])
```

```
count     3.0
mean     42.0
std      42.0
min       0.0
10%       8.4
44%      36.96
50%      42.0
67%      56.28
max      84.0
dtype: Float64
```

Binning algorithms

Binning is the process of taking a continuous variable and categorizing it into discrete buckets. It can be useful to turn a potentially infinite amount of values into a finite amount of "bins" for your analysis.

How to do it

Let's imagine we have collected survey data from users of a system. One of the survey questions asks users for their age, producing data that looks like:

```python
df = pd.DataFrame([
    ["Jane", 34],
    ["John", 18],
    ["Jamie", 22],
    ["Jessica", 36],
    ["Jackie", 33],
    ["Steve", 40],
    ["Sam", 30],
    ["Stephanie", 66],
    ["Sarah", 55],
    ["Aaron", 22],
    ["Erin", 28],
    ["Elsa", 37],
], columns=["name", "age"])
df = df.convert_dtypes(dtype_backend="numpy_nullable")
df.head()
```

	name	age
0	Jane	34
1	John	18
2	Jamie	22
3	Jessica	36
4	Jackie	33

Rather than treating each age as an individual number, we will use pd.cut to place each record into an age group. As a first attempt, let's pass our pd.Series and the number of bins we would like to generate as arguments:

```python
pd.cut(df["age"], 4)
```

0	(30.0, 42.0]
1	(17.952, 30.0]
2	(17.952, 30.0]
3	(30.0, 42.0]
4	(30.0, 42.0]
5	(30.0, 42.0]
6	(17.952, 30.0]
7	(54.0, 66.0]
8	(54.0, 66.0]
9	(17.952, 30.0]

```
10     (17.952, 30.0]
11       (30.0, 42.0]
Name: age, dtype: category
Categories (4, interval[float64, right]): [(17.952, 30.0] < (30.0, 42.0] <
(42.0, 54.0] < (54.0, 66.0]]
```

This produces a pd.CategoricalDtype with 4 distinct intervals – (17.952, 30.0], (30.0, 42.0], (42.0, 54.0], and (54.0, 66.0]. Save some unexpected decimal places on the first bin, which starts at 17.952, these bins all cover an equidistant range of 12 years, which was derived from the fact that the maximum value (66) minus the lowest value (18) yields a total age gap of 48 years, which, when divided equally by 4, gives us the 12-year range for each bin.

The age 17.952 we see in the first bin may make sense to pandas internally for whatever algorithm it chose to determine the buckets, but it is ultimately uninteresting to us since we know we are dealing with whole numbers. Fortunately, this can be controlled via the precision= keyword argument to remove any decimal places:

```
pd.cut(df["age"], 4, precision=0)
```

```
0      (30.0, 42.0]
1      (18.0, 30.0]
2      (18.0, 30.0]
3      (30.0, 42.0]
4      (30.0, 42.0]
5      (30.0, 42.0]
6      (18.0, 30.0]
7      (54.0, 66.0]
8      (54.0, 66.0]
9      (18.0, 30.0]
10     (18.0, 30.0]
11     (30.0, 42.0]
Name: age, dtype: category
Categories (4, interval[float64, right]): [(18.0, 30.0] < (30.0, 42.0] < (42.0,
54.0] < (54.0, 66.0]]
```

pd.cut does not limit us to producing equally sized bins like this. If, instead, we wanted to place each person into 10-year age buckets, we could provide those ranges as the second argument:

```
pd.cut(df["age"], [10, 20, 30, 40, 50, 60, 70])
```

```
0      (30, 40]
1      (10, 20]
2      (20, 30]
3      (30, 40]
4      (30, 40]
5      (30, 40]
```

```
6        (20, 30]
7        (60, 70]
8        (50, 60]
9        (20, 30]
10       (20, 30]
11       (30, 40]
Name: age, dtype: category
Categories (6, interval[int64, right]): [(10, 20] < (20, 30] < (30, 40] < (40,
50] < (50, 60] < (60, 70]]
```

However, this is a little too strict because it would not account for users over the age of 70. To handle that, we could change our last bin edge from 70 to 999 and treat it as a catch-all:

```
pd.cut(df["age"], [10, 20, 30, 40, 50, 60, 999])
```

```
0        (30, 40]
1        (10, 20]
2        (20, 30]
3        (30, 40]
4        (30, 40]
5        (30, 40]
6        (20, 30]
7        (60, 999]
8        (50, 60]
9        (20, 30]
10       (20, 30]
11       (30, 40]
Name: age, dtype: category
Categories (6, interval[int64, right]): [(10, 20] < (20, 30] < (30, 40] < (40,
50] < (50, 60] < (60, 999]]
```

In turn, this produced a label of (60, 999), which leaves something to be desired from a display perspective. If we are not happy with the default labels produced, we can control their output with the labels= argument:

```
pd.cut(
    df["age"],
    [10, 20, 30, 40, 50, 60, 999],
    labels=["10-20", "20-30", "30-40", "40-50", "50-60", "60+"],
)
```

```
0        30-40
1        10-20
2        20-30
3        30-40
```

```
 4     30-40
 5     30-40
 6     20-30
 7       60+
 8     50-60
 9     20-30
10     20-30
11     30-40
Name: age, dtype: category
Categories (6, object): ['10-20' < '20-30' < '30-40' < '40-50' < '50-60' <
'60+']
```

However, our labels above are not *quite* right. Note that we provided both `30-40` and `40-50`, but what happens if someone is exactly 40 years old? What bin are they placed in?

Fortunately, we can see this in our data already from `Steve`, who perfectly matches this criteria. If you inspect the default bin he is placed in, it appears as `(30, 40]`:

```
df.assign(age_bin=lambda x: pd.cut(x["age"], [10, 20, 30, 40, 50, 60, 999]))
```

	name	age	age_bin
0	Jane	34	(30, 40]
1	John	18	(10, 20]
2	Jamie	22	(20, 30]
3	Jessica	36	(30, 40]
4	Jackie	33	(30, 40]
5	Steve	40	(30, 40]
6	Sam	30	(20, 30]
7	Stephanie	66	(60, 999]
8	Sarah	55	(50, 60]
9	Aaron	22	(20, 30]
10	Erin	28	(20, 30]
11	Elsa	37	(30, 40]

Binning, by default, is *right inclusive*, meaning each bin can be thought of as *up to and including* a particular value. If we wanted behavior that was *up to but not including*, we could control this with the `right` argument:

```
df.assign(
    age_bin=lambda x: pd.cut(x["age"], [10, 20, 30, 40, 50, 60, 999],
right=False)
)
```

	name	age	age_bin
0	Jane	34	[30, 40)
1	John	18	[10, 20)

```
2         Jamie      22    [20, 30)
3         Jessica    36    [30, 40)
4         Jackie     33    [30, 40)
5         Steve      40    [40, 50)
6         Sam        30    [30, 40)
7         Stephanie  66    [60, 999)
8         Sarah      55    [50, 60)
9         Aaron      22    [20, 30)
10        Erin       28    [20, 30)
11        Elsa       37    [30, 40)
```

This changed the bin for Steve from (30, 40] to [40, 50). In the default string representation, the square bracket signifies the edge being *inclusive* of a particular value, whereas the parenthesis is *exclusive*.

One-hot encoding with pd.get_dummies

It is not uncommon in data analysis and machine learning applications to take data that is categorical in nature and convert it into a sequence of 0/1 values, as the latter can be more easily interpreted by numeric algorithms. This process is often called *one-hot encoding*, and the outputs are typically referred to as *dummy indicators*.

How to do it

Let's start with a small pd.Series containing a discrete set of colors:

```python
ser = pd.Series([
    "green",
    "brown",
    "blue",
    "amber",
    "hazel",
    "amber",
    "green",
    "blue",
    "green",
], name="eye_colors", dtype=pd.StringDtype())
ser
```

```
0     green
1     brown
2      blue
3     amber
4     hazel
```

```
5      amber
6      green
7       blue
8      green
Name: eye_colors, dtype: string
```

Passing this as an argument to `pd.get_dummies` will create a like-indexed `pd.DataFrame` with a Boolean column for each color. Each row has one column with `True` that maps it back to its original value; all other columns in the same row will be `False`:

```
pd.get_dummies(ser)
```

	amber	blue	brown	green	hazel
0	False	False	False	True	False
1	False	False	True	False	False
2	False	True	False	False	False
3	True	False	False	False	False
4	False	False	False	False	True
5	True	False	False	False	False
6	False	False	False	True	False
7	False	True	False	False	False
8	False	False	False	True	False

If we are not satisfied with the default column names, we can modify them by adding a prefix. A common convention in data modeling is to prefix a Boolean column with `is_`:

```
pd.get_dummies(ser, prefix="is")
```

	is_amber	is_blue	is_brown	is_green	is_hazel
0	False	False	False	True	False
1	False	False	True	False	False
2	False	True	False	False	False
3	True	False	False	False	False
4	False	False	False	False	True
5	True	False	False	False	False
6	False	False	False	True	False
7	False	True	False	False	False
8	False	False	False	True	False

Chaining with .pipe

When writing pandas code, there are two major stylistic forms that developers follow. The first approach makes liberal use of variables throughout a program, whether that means creating new variables like:

```
df = pd.DataFrame(...)
df1 = do_something(df)
```

```
df2 = do_another_thing(df1)
df3 = do_yet_another_thing(df2)
```

or simply reassigning to the same variable repeatedly:

```
df = pd.DataFrame(...)
df = do_something(df)
df = do_another_thing(df)
df = do_yet_another_thing(df)
```

The alternative approach is to express your code as a *pipeline*, where each step accepts and returns a
pd.DataFrame:

```
(
    pd.DataFrame(...)
    .pipe(do_something)
    .pipe(do_another_thing)
    .pipe(do_yet_another_thing)
)
```

With the variable-based approach, you must create multiple variables in your program, or change the
state of a pd.DataFrame at every reassignment. The pipeline approach, by contrast, does not create
any new variables, nor does it change the state of your pd.DataFrame.

While the pipeline approach could theoretically be better handled by a query optimizer, pandas does
not offer such a feature as of the time of writing, and it is hard to guess what that may look like in the
future. As such, the choice between the two approaches makes almost no difference for performance;
it is truly a matter of style.

I encourage you to familiarize yourself with both approaches. You may at times find it easier to express
your code as a pipeline; at other times, that may feel burdensome. There is no hard requirement to
use one or the other, so you can mix and match the styles freely throughout your code.

How to do it

Let's start with a very basic pd.DataFrame. The columns and their contents are not important for now:

```
df = pd.DataFrame({
    "col1": pd.Series([1, 2, 3], dtype=pd.Int64Dtype()),
    "col2": pd.Series(["a", "b", "c"], dtype=pd.StringDtype()),
})
df
```

```
    col1    col2
0   1       a
1   2       b
2   3       c
```

Now let's create some sample functions that will change the content of the columns. These functions should accept and return a pd.DataFrame, which you can see from the code annotations:

```python
def change_col1(df: pd.DataFrame) -> pd.DataFrame:
    return df.assign(col1=pd.Series([4, 5, 6], dtype=pd.Int64Dtype()))


def change_col2(df: pd.DataFrame) -> pd.DataFrame:
    return df.assign(col2=pd.Series(["X", "Y", "Z"], dtype=pd.StringDtype()))
```

As mentioned in the introduction to this recipe, one of the most common ways to apply these functions would be to list them out as separate steps in our program, assigning the results of each step to a new variable along the way:

```python
df2 = change_col1(df)
df3 = change_col2(df2)
df3
```

	col1	col2
0	4	X
1	5	Y
2	6	Z

If we wanted to avoid the use of intermediate variables altogether, we could have also tried to nest the function calls inside of one another:

```python
change_col2(change_col1(df))
```

	col1	col2
0	4	X
1	5	Y
2	6	Z

However, that doesn't make the code any more readable, especially given the fact that change_col1 is executed before change_col2.

By expressing this as a pipeline, we can avoid the use of variables and more easily express the order of operations being applied. To achieve this, we are going to reach for the pd.DataFrame.pipe method:

```python
df.pipe(change_col1).pipe(change_col2)
```

	col1	col2
0	4	X
1	5	Y
2	6	Z

As you can see, we have gotten back the same result as before, but without the use of variables and in a way that is arguably more readable.

In case any of the functions you want to apply in a pipeline need to accept more arguments, `pd.DataFrame.pipe` is able to forward them along for you. For instance, let's see what happens if we add a new `str_case` parameter to our `change_col2` function:

```python
from typing import Literal


def change_col2(
        df: pd.DataFrame,
        str_case: Literal["upper", "lower"]
) -> pd.DataFrame:
    if str_case == "upper":
        values = ["X", "Y", "Z"]
    else:
        values = ["x", "y", "z"]

    return df.assign(col2=pd.Series(values, dtype=pd.StringDtype()))
```

As you can see with `pd.DataFrame.pipe`, you can simply pass that argument along as either a positional or keyword argument, just as if you were invoking the `change_col2` function directly:

```python
df.pipe(change_col2, str_case="lower")
```

```
     col1    col2
0    1       x
1    2       y
2    3       z
```

To reiterate what we mentioned in the introduction to this recipe, there is little to no functional difference between these styles. I encourage you to learn them both as you will inevitably see code written both ways. For your own development, you may even find that mixing and matching the approaches works best.

Selecting the lowest-budget movies from the top 100

Now that we have covered many of the core pandas algorithms from a theoretical level, we can start looking at more "real world" datasets and touch on common ways to explore them.

Top N analysis is a common technique whereby you filter your data based on how your data performs when measured by a single variable. Most analytics tools have the capability to help you filter your data to answer questions like *What are the top 10 customers by sales?* or, *What are the 10 products with the lowest inventory?*. When chained together, you can even form catchy news headlines such as *Out of the Top 100 Universities, These 5 Have the Lowest Tuition Fees*, or *From the Top 50 Cities to Live, These 10 Are the Most Affordable*.

Given how common these types of analyses are, pandas offers built-in functionality to help you easily perform them. In this recipe, we will take a look at `pd.DataFrame.nlargest` and `pd.DataFrame.nsmallest` and see how we can use them together to answer a question like *From the top 100 movies, which had the lowest budget?*.

How to do it

Let's start by reading in the movie dataset and selecting the columns `movie_title`, `imdb_score`, `budget`, and `gross`:

```
df = pd.read_csv(
    "data/movie.csv",
    usecols=["movie_title", "imdb_score", "budget", "gross"],
    dtype_backend="numpy_nullable",
)
df.head()
```

```
       gross        movie_title                                 budget       imdb_
score
0      760505847.0  Avatar                                      237000000.0  7.9
1      309404152.0  Pirates of the Caribbean: At World's End    300000000.0
7.1
2      200074175.0  Spectre                                     245000000.0  6.8
3      448130642.0  The Dark Knight Rises                       250000000.0  8.5
4      <NA>         Star Wars: Episode VII - The Force Awakens   <NA>         7.1
```

The `pd.DataFrame.nlargest` method can be used to select the top 100 movies by `imdb_score`:

```
df.nlargest(100, "imdb_score").head()
```

```
       gross        movie_title               budget       imdb_score
2725   <NA>         Towering Inferno          <NA>         9.5
1920   28341469.0   The Shawshank Redemption  25000000.0   9.3
3402   134821952.0  The Godfather             6000000.0    9.2
2779   447093.0     Dekalog                   <NA>         9.1
4312   <NA>         Kickboxer: Vengeance      17000000.0   9.1
```

Now that we have the top 100 selected, we can chain in a call to `pd.DataFrame.nsmallest` to return the five lowest-budget movies among those:

```
df.nlargest(100, "imdb_score").nsmallest(5, "budget")
```

```
       gross       movie_title        budget      imdb_score
4804   <NA>        Butterfly Girl     180000.0    8.7
4801   925402.0   Children of Heaven  180000.0    8.5
4706   <NA>        12 Angry Men       350000.0    8.9
```

```
4550      7098492.0    A Separation                500000.0    8.4
4636      133778.0     The Other Dream Team        500000.0    8.4
```

There's more...

It is possible to pass a list of column names as the columns= parameter of the pd.DataFrame.nlargest and pd.DataFrame.nsmallest methods. This would only be useful to break ties in the event that there were duplicate values sharing the *nth* ranked spot in the first column in the list.

To see where this matters, let's try to just select the top 10 movies by imdb_score:

```
df.nlargest(10, "imdb_score")
```

```
          gross     movie_title       budget    imdb_score
2725      <NA>      Towering Inferno        <NA>      9.5
1920      28341469.0        The Shawshank Redemption          25000000.0      9.3
3402      134821952.0       The Godfather    6000000.0        9.2
2779      447093.0          Dekalog     <NA>      9.1
4312      <NA>      Kickboxer: Vengeance    17000000.0      9.1
66        533316061.0       The Dark Knight 185000000.0      9.0
2791      57300000.0        The Godfather: Part II       13000000.0      9.0
3415      <NA>      Fargo      <NA>      9.0
335       377019252.0       The Lord of the Rings: The Return of the King
94000000.0        8.9
1857      96067179.0        Schindler's List          22000000.0      8.9
```

As you can see, the lowest imdb_score from the top 10 is 8.9. However, there are more than 10 movies that have a score of 8.9 and above:

```
df[df["imdb_score"] >= 8.9]
```

```
          gross     movie_title       budget    imdb_score
66        533316061.0       The Dark Knight 185000000.0      9.0
335       377019252.0       The Lord of the Rings: The Return of the King
94000000.0        8.9
1857      96067179.0        Schindler's List          22000000.0      8.9
1920      28341469.0        The Shawshank Redemption          25000000.0      9.3
2725      <NA>      Towering Inferno        <NA>      9.5
2779      447093.0          Dekalog <NA>      9.1
2791      57300000.0        The Godfather: Part II       13000000.0      9.0
3295      107930000.0       Pulp Fiction    8000000.0        8.9
3402      134821952.0       The Godfather    6000000.0        9.2
3415      <NA>      Fargo      <NA>      9.0
4312      <NA>      Kickboxer: Vengeance    17000000.0      9.1
4397      6100000.0         The Good, the Bad and the Ugly    1200000.0        8.9
4706      <NA>      12 Angry Men        350000.0        8.9
```

The movies that were a part of the top 10 just happened to be the first two movies pandas came across with that score. However, you can use the gross column as the tiebreaker:

```
df.nlargest(10, ["imdb_score", "gross"])
```

```
         gross    movie_title        budget   imdb_score
2725     <NA>     Towering Inferno          <NA>     9.5
1920     28341469.0      The Shawshank Redemption      25000000.0      9.3
3402     134821952.0     The Godfather   6000000.0     9.2
2779     447093.0        Dekalog    <NA>    9.1
4312     <NA>     Kickboxer: Vengeance    17000000.0      9.1
66       533316061.0     The Dark Knight    185000000.0      9.0
2791     57300000.0      The Godfather: Part II     13000000.0       9.0
3415     <NA>     Fargo    <NA>    9.0
335      377019252.0     The Lord of the Rings: The Return of the King
94000000.0       8.9
3295     107930000.0     Pulp Fiction    8000000.0      8.9
```

With that, you see that Pulp Fiction replaced Schindler's List in our top 10 analysis, given that it grossed higher.

Calculating a trailing stop order price

There are many strategies to trade stocks. One basic type of trade that many investors employ is the *stop order*. A stop order is an order placed by an investor to buy or sell a stock that executes whenever the market price reaches a certain point. Stop orders are useful to both prevent huge losses and protect gains.

In a typical stop order, the price does not change throughout the lifetime of the order. For instance, if you purchased a stock for $100 per share, you might want to set a stop order at $90 per share to limit your downside to 10%.

A more advanced strategy would be to continually modify the sale price of the stop order to track the value of the stock if it increases in value. This is called a *trailing stop order*. Concretely, if the same $100 stock increases to $120, then a trailing stop order 10% below the current market value would move the sale price to $108.

The trailing stop order never moves down and is always tied to the maximum value since the time of purchase. If the stock fell from $120 to $110, the stop order would still remain at $108. It would only increase if the price moved above $120.

This recipe determines the trailing stop order price given an initial purchase price for any stock using the `pd.Series.cummax` method and how `pd.Series.cummin` could instead be used to handle short positions. We will also see how the `pd.Series.idxmax` method can be used to identify the day the stop order would have been triggered.

How to do it

To get started, we will work with Nvidia (NVDA) stock and assume a purchase on the first trading day of 2020:

```python
df = pd.read_csv(
    "data/NVDA.csv",
    usecols=["Date", "Close"],
    parse_dates=["Date"],
    index_col=["Date"],
    dtype_backend="numpy_nullable",
)
df.head()
```

```
ValueError: not all elements from date_cols are numpy arrays
```

In the pandas 2.2 series, there is a bug that prevents the preceding code block from running, instead throwing a `ValueError`. If affected by this bug, you can alternatively run `pd.read_csv` without the `dtype_backend` argument, and add in a call to `pd.DataFrame.convert_dtypes` instead:

```python
df = pd.read_csv(
    "data/NVDA.csv",
    usecols=["Date", "Close"],
    parse_dates=["Date"],
    index_col=["Date"],
).convert_dtypes(dtype_backend="numpy_nullable")

df.head()
```

```
                      Close
Date
2020-01-02       59.977501
2020-01-03       59.017502
2020-01-06       59.264999
2020-01-07       59.982498
2020-01-08       60.095001
```

For more information, see pandas bug issue #57930 (https://github.com/pandas-dev/pandas/issues/57930).

Regardless of which path you took, be aware that `pd.read_csv` returns a `pd.DataFrame`, but for this analysis we will only need a `pd.Series`. To perform that conversion, you can call `pd.DataFrame.squeeze`, which will reduce the object from two to one dimension, if possible:

```python
ser = df.squeeze()
ser.head()
```

```
Date
2020-01-02    59.977501
2020-01-03    59.017502
2020-01-06    59.264999
2020-01-07    59.982498
2020-01-08    60.095001
Name: Close, dtype: float64
```

With that, we can use the `pd.Series.cummax` method to track the highest closing price seen to date:

```
ser_cummax = ser.cummax()
ser_cummax.head()
```

```
Date
2020-01-02    59.977501
2020-01-03    59.977501
2020-01-06    59.977501
2020-01-07    59.982498
2020-01-08    60.095001
Name: Close, dtype: float64
```

To create a trailing stop order that limits our downside to 10%, we can chain in a multiplication by `0.9`:

```
ser.cummax().mul(0.9).head()
```

```
Date
2020-01-02    53.979751
2020-01-03    53.979751
2020-01-06    53.979751
2020-01-07    53.984248
2020-01-08    54.085501
Name: Close, dtype: float64
```

The `pd.Series.cummax` method works by retaining the maximum value encountered up to and including the current value. Multiplying this series by 0.9, or whatever cushion you would like to use, creates the trailing stop order. In this particular example, NVDA increased in value, and thus, its trailing stop has also increased.

On the flip side, let's say we were pessimistic about NVDA stock during this timeframe, and we wanted to short the stock. However, we still wanted to put a stop order in place to limit the downside to a 10% increase in value.

For this, we can simply replace our usage of `pd.Series.cummax` with `pd.Series.cummin` and multiply by `1.1` instead of `0.9`:

```
ser.cummin().mul(1.1).head()
```

```
Date
2020-01-02    65.975251
2020-01-03    64.919252
2020-01-06    64.919252
2020-01-07    64.919252
2020-01-08    64.919252
Name: Close, dtype: float64
```

There's more...

With our trailing stop orders calculated, we can easily determine the days where we would have fallen off of the cumulative maximum by more than our threshold:

```
stop_prices = ser.cummax().mul(0.9)
ser[ser <= stop_prices]
```

```
Date
2020-02-24    68.320000
2020-02-25    65.512497
2020-02-26    66.912498
2020-02-27    63.150002
2020-02-28    67.517502
                 ...
2023-10-27   405.000000
2023-10-30   411.609985
2023-10-31   407.799988
2023-11-01   423.250000
2023-11-02   435.059998
Name: Close, Length: 495, dtype: float64
```

If we only cared to identify the very first day where we fell below the cumulative maximum, we could use the pd.Series.idxmax method. This method works by first calculating the maximum value within a pd.Series, and then returns the first-row index where that maximum was encountered:

```
(ser <= stop_prices).idxmax()
```

```
Timestamp('2020-02-24 00:00:00')
```

The expression ser <= stop_prices gives back a Boolean pd.Series containing True=/=False values, with each True record indicating where the stock price is at or below the stop price we already calculated. pd.Series.idxmax will consider True to be the maximum value in that pd.Series; so, by returning the first index label where True was seen as a value, it tells us the first day that our trailing stop order should have been triggered.

This recipe gives us just a taste of how useful pandas may be for trading securities.

Finding the baseball players best at...

The American sport of baseball has long been a subject of intense analytical research, with data collection dating back to the early 1900s. For Major League baseball teams, advanced data analysis helps answer questions like *How much should I pay for X player?* and *What should I do in the game given the current state of things?*, For fans, that same data can be used as fodder for endless debates around *who is the greatest player ever*.

For this recipe, we are going to use data that was collected from `retrosheet.org`. Per the Retrosheet licensing requirements, you should be aware of the following legal disclaimer:

 The information used here was obtained free of charge and is copyrighted by Retrosheet. Interested parties may contact Retrosheet at `www.retrosheet.org`.

From its raw form, the data was summarized to show the common baseball metrics for at bat (ab), hits (h), runs scored (r), and home runs (hr) for professional players in the years 2020–2023.

How to do it

Let's start by reading in our summarized data and setting the `id` column (which represents a unique player) as the index:

```
df = pd.read_parquet(
    "data/mlb_batting_summaries.parquet",
).set_index("id")

df
```

	ab	r	h	hr
id				
abadf001	0	0	0	0
abboa001	0	0	0	0
abboc001	3	0	1	0
abrac001	847	116	208	20
abrea001	0	0	0	0
...	...	...	...	...
zimmk001	0	0	0	0
zimmr001	255	27	62	14
zubet001	1	0	0	0
zunig001	0	0	0	0
zunim001	572	82	111	41

2183 rows × 4 columns

In baseball, it is rather rare for a player to dominate all statistical categories. Oftentimes, a player with a lot of home runs (hr) will be more powerful and can hit the ball farther, but may do so less frequently than a player more specialized to collect a lot of hits (h). With pandas, we are fortunate to not have to dive into each metric individually; a simple call to pd.DataFrame.idxmax will look at each column, find the maximum value, and return the row index value associated with that maximum value for you:

```
df.idxmax()
```

```
ab       semim001
r        freef001
h        freef001
hr       judga001
dtype: string
```

As you can see, player semim001 (Marcus Semien) had the most at bats, freef001 (Freddie Freeman) had the most runs and hits, and judga001 (Aaron Judge) hit the most home runs in this timeframe.

If you wanted to look deeper into how these great players performed across all categories, you could take the output of pd.DataFrame.idxmax, subsequently call pd.Series.unique on the values, and use that as a mask for the overall pd.DataFrame:

```
best_players = df.idxmax().unique()
mask = df.index.isin(best_players)
df[mask]
```

```
              ab      r      h      hr
id
freef001    1849    368    590    81
judga001    1487    301    433    138
semim001    1979    338    521    100
```

There's more...

For a nice visual enhancement to this data, you can use pd.DataFrame.style.highlight_max to very specifically show which category these players were the best at:

```
df[mask].style.highlight_max()
```

id	ab	r	h	hr
freef001	1849	368	590	81
judga001	1487	301	433	138
semim001	1979	338	521	100

Figure 5.1: Jupyter Notebook output of a DataFrame highlighting max value per column

Understanding which position scores the most per tea

In baseball, teams are allowed 9 batters in a "lineup," with 1 representing the first person to bat and 9 representing the last. Over the course of a game, teams cycle through batters in order, starting over with the first batter after the last has batted.

Typically, teams place some of their best hitters toward the "top of the lineup" (i.e., lower number positions) to maximize the opportunity for them to come around and score. However, this does not always mean that the person who bats in position 1 will always be the first to score.

In this recipe, we are going to look at all Major League baseball teams from 2000–2023 and find the position that scored the most runs for a team over each season.

How to do it

Much like we did in the *Finding the baseball players best at...* recipe, we are going to use data taken from `retrosheet.org`. For this particular dataset, we are going to set the `year` and `team` columns in the row index, leaving the remaining columns to show the position in the batting order:

```
df = pd.read_parquet(
    "data/runs_scored_by_team.parquet",
).set_index(["year", "team"])

df
```

		1	2	3	...	7	8	9
year	team							
2000	ANA	124	107	100	...	77	76	54
	ARI	110	106	109	...	72	68	40
	ATL	113	125	124	...	77	74	39
	BAL	106	106	92	...	83	78	74
	BOS	99	107	99	...	75	66	62
...	...	...	...	...	...	...	...	...
2023	SLN	105	91	85	...	70	55	74
	TBA	121	120	93	...	78	95	98
	TEX	126	115	91	...	80	87	81
	TOR	91	97	85	...	64	70	79
	WAS	110	90	87	...	63	67	64

720 rows × 9 columns

With `pd.DataFrame.idxmax`, we can see for every year and team which position scored the most runs. However, with this dataset, the index label we would like `pd.DataFrame.idxmax` to identify is actually in the columns and not the rows. Fortunately, pandas can still calculate this easily with the `axis=1` argument:

```
df.idxmax(axis=1)
```

```
year  team
2000  ANA      1
      ARI      1
      ATL      2
      BAL      1
      BOS      4
             ..
2023  SLN      1
      TBA      1
      TEX      1
      TOR      2
      WAS      1
Length: 720, dtype: object
```

From there, we can use pd.Series.value_counts to understand the number of times a given position in the order represented the most runs scored for a team. We are also going to use the normalize=True argument, which will give us a frequency instead of a total:

```
df.idxmax(axis=1).value_counts(normalize=True)
```

```
1    0.480556
2    0.208333
3    0.202778
4    0.088889
5    0.018056
6    0.001389
Name: proportion, dtype: float64
```

Unsurprisingly, the first batter scored most frequently accounted for the most runs, doing so for 48% of the teams.

There's more...

We might want to explore more and answer the question: *For teams where the first batter scored the most runs, who scored the second-most?*

To calculate this, we can create a mask to filter on teams where the first batter scored the most, drop that column from our dataset, and then repeat with the same pd.DataFrame.idxmax approach to identify the position next in line:

```
mask = df.idxmax(axis=1).eq("1")
df[mask].drop(columns=["1"]).idxmax(axis=1).value_counts(normalize=True)
```

```
2    0.497110
3    0.280347
4    0.164740
```

```
5      0.043353
6      0.014451
Name: proportion, dtype: float64
```

As you can see, if a team's first batter does not lead the team in runs scored, the second batter ends up being the leader almost 50% of the time.

Join our community on Discord

Join our community's Discord space for discussions with the authors and other readers:

`https://packt.link/pandas`

6

Visualization

Visualization is a critical component in exploratory data analysis, as well as presentations and applications. During exploratory data analysis, you are usually working alone or in small groups and need to create plots quickly to help you better understand your data. Visualizations can help you identify outliers and missing data, or they can spark other questions of interest that will lead to further analysis and more visualizations. This type of visualization is usually not done with the end user in mind. It is strictly to help you better your current understanding. The plots do not have to be perfect.

When preparing visualizations for a report or application, a different approach must be used. You should pay attention to small details. Also, you usually will have to narrow down all possible visualizations to only the select few that best represent your data. Good data visualizations have the viewer enjoying the experience of extracting information. Almost like movies that viewers can get lost in, good visualizations will have lots of information that really sparks interest.

Out of the box, pandas has the `pd.Series.plot` and `pd.DataFrame.plot` methods to help you quickly generate plots. These methods dispatch to a *plotting backend*, which by default is Matplotlib (`https://matplotlib.org/`).

We will discuss different backends later in this chapter, but for now, let's start by installing Matplotlib and PyQt5, which Matplotlib uses to draw plots:

```
python -m pip install matplotlib pyqt5
```

All code samples in this chapter are assumed to be preceded by the following import:

```
import matplotlib.pyplot as plt
plt.ion()
```

The previous command enables Matplotlib's *interactive mode*, which will create and update your plots automatically every time a plotting command is executed. If, for whatever reason, you run a plotting command but no plot appears, you likely are in non-interactive mode (you can check this with `matplotlib.pyplot.isinteractive()`), and you will need to explicitly call `matplotlib.pyplot.show()` to make your plots appear.

We are going to cover the following recipes in this chapter:

- Creating charts from aggregated data
- Plotting distributions of non-aggregated data
- Further plot customization with Matplotlib
- Exploring scatter plots
- Exploring categorical data
- Exploring continuous data
- Using seaborn for advanced plots

Creating charts from aggregated data

The pandas library makes it easy to visualize data in `pd.Series` and `pd.DataFrame` objects, using the `pd.Series.plot` and `pd.DataFrame.plot` methods, respectively. In this recipe we are going to start with relatively basic line, bar, area, and pie charts, while also seeing the high-level customization options pandas offers. While, these chart types are simple, using them effectively can be immensely helpful to explore your data, identify trends, and share your research with non-technical associates.

It is important to note that these chart types expect your data to already be aggregated, which our sample data in this recipe will reflect. If you are working with data that is not yet aggregated, you will need to use techniques that you will encounter in *Chapter 7, Reshaping DataFrames*, and *Chapter 8, Group By*, or use the techniques shown in the *Using Seaborn for advanced plots* recipe later in this chapter.

How to do it

Let's create a simple `pd.Series` showing book sales over the course of a 7-day period. We are intentionally going to use row index labels of the form *Day n*, which will provide a good visual clue on the different chart types we create:

```
ser = pd.Series(
    (x ** 2 for x in range(7)),
    name="book_sales",
    index=(f"Day {x + 1}" for x in range(7)),
    dtype=pd.Int64Dtype(),
)

ser
```

```
Day 1    0
Day 2    1
Day 3    4
Day 4    9
Day 5    16
```

```
Day 6      25
Day 7      36
Name: book_sales, dtype: Int64
```

A call to pd.Series.plot without any arguments will produce a line chart, where the labels used on the x-axis come from the row index and the values on the Y-axis correspond to the data within the pd.Series:

```
ser.plot()
```

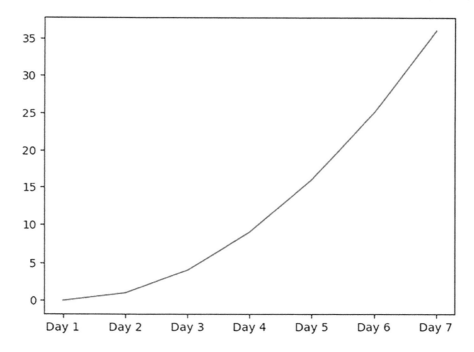

A line chart treats our data as if it is completely continuous, yielding a visualization that appears to show values in between each day, even though that does not exist in our data. A better visualization for our pd.Series would be a bar chart that displays each day discretely, which we can get just by passing the kind="bar" argument to the pd.Series.plot method:

```
ser.plot(kind="bar")
```

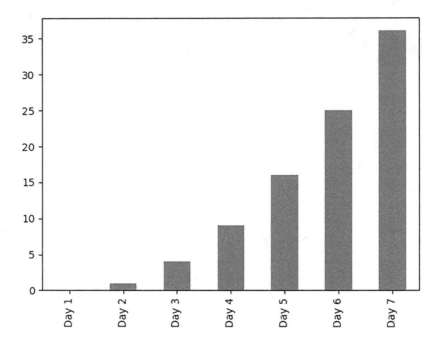

Once again, the row index labels appear on the *X*-axis and the values appear on the *Y*-axis. This helps you read the visualization from left to right, but in some circumstances, you may find it easier to read values from top to bottom. In pandas, such a visualization would be considered a *horizontal bar chart*, which can be rendered by using the kind="barh" argument:

```
ser.plot(kind="barh")
```

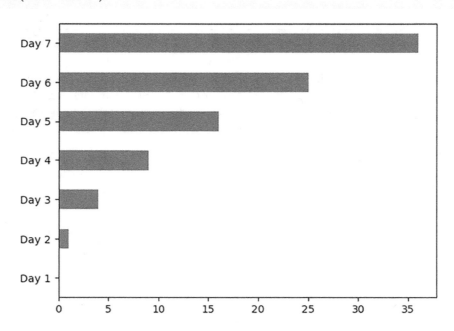

A kind="area" argument will produce an area chart, which is like a line chart but fills in the area underneath the line:

```
ser.plot(kind="area")
```

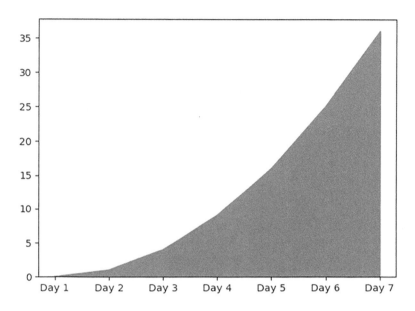

Last but not least, we have pie charts. Unlike all of the visualizations introduced so far, a pie chart does not have both an x- and a y-axis. Instead, each label from the row index represents a different slice of the pie, whose size is dictated by the associated value in our pd.Series:

```
ser.plot(kind="pie")
```

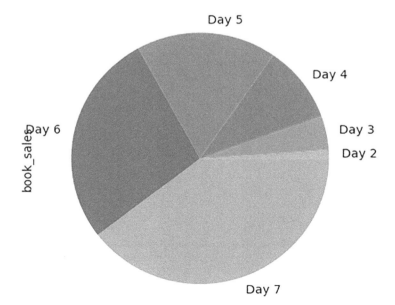

When working with a pd.DataFrame, the API through which you generate charts stays consistent, although you may find that you need to provide more keyword arguments to get the desired visualization.

To see this in action, let's extend our data to show both book_sales and book_returns:

```python
df = pd.DataFrame({
    "book_sales": (x ** 2 for x in range(7)),
    "book_returns": [3, 2, 1, 0, 1, 2, 3],
}, index=(f"Day {x + 1}" for x in range(7)))
df = df.convert_dtypes(dtype_backend="numpy_nullable")
df
```

	book_sales	book_returns
Day 1	0	3
Day 2	1	2
Day 3	4	1
Day 4	9	0
Day 5	16	1
Day 6	25	2
Day 7	36	3

Just like we saw with pd.Series.plot, the default call to pd.DataFrame.plot will give us a line plot, with each column represented by its own line:

```python
df.plot()
```

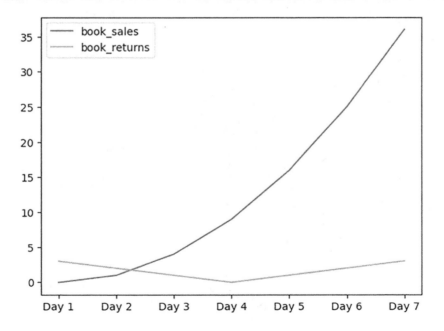

Once again, to turn this into a bar chart, you would just need to pass `kind="bar"` to the plotting method:

```
df.plot(kind="bar")
```

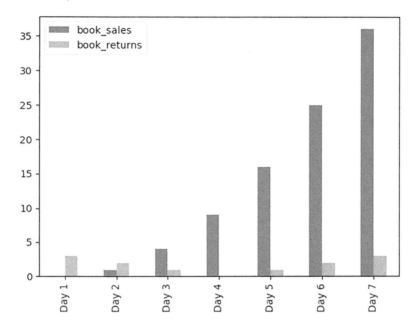

By default, pandas will present each column as a separate bar on the chart. If you wanted instead to stack the columns on top of one another, pass `stacked=True`:

```
df.plot(kind="bar", stacked=True)
```

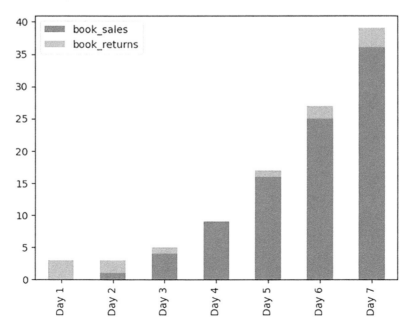

The same behavior can be seen with a horizontal bar chart. By default, the columns will not be stacked:

```
df.plot(kind="barh")
```

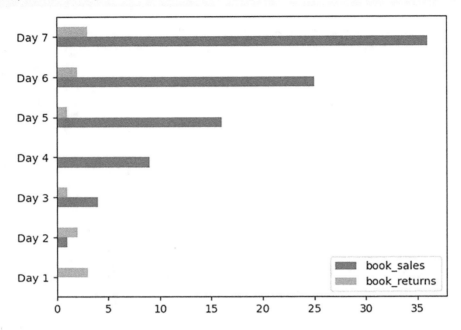

But passing `stacked=True` will place the bars on top of one another:

```
df.plot(kind="barh", stacked=True)
```

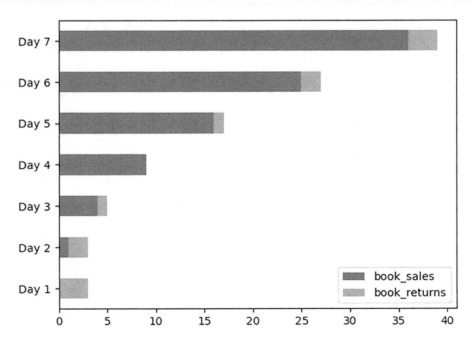

When using a pd.DataFrame with an area chart, the default behavior is to stack the columns:

```
df.plot(kind="area")
```

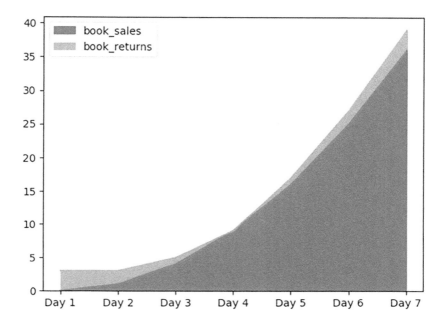

To unstack, pass stacked=False and include an alpha= argument to introduce transparency. The value of this argument should be between 0 and 1, with values closer to 0 making the chart more transparent:

```
df.plot(kind="area", stacked=False, alpha=0.5)
```

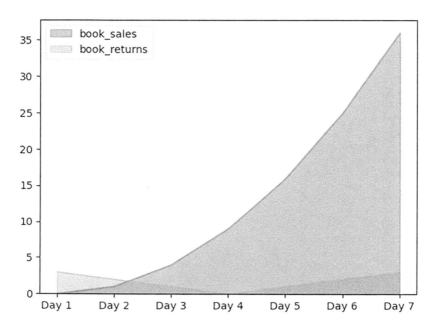

There's more...

The examples in this recipe used the minimum amount of arguments to produce visuals. However, the plotting methods accept many more arguments to control things like titles, labels, colors, etc.

If you want to add a title to your visualization, simply pass it as the `title=` argument:

```
ser.plot(
    kind="bar",
    title="Book Sales by Day",
)
```

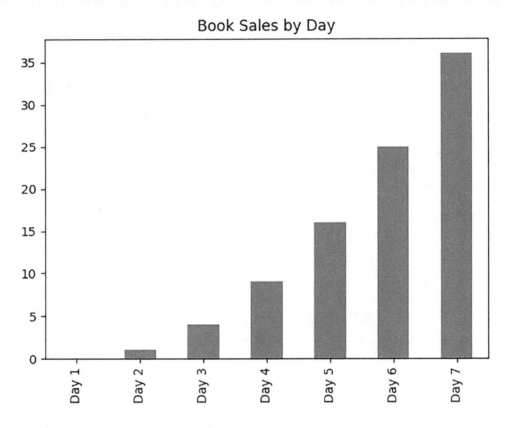

The color= argument can be used to change the color of the lines, bars, and markers in your chart. Color can be expressed using RGB hex codes (like #00008B for dark blue) or by using a Matplotlib named color like seagreen (https://matplotlib.org/stable/gallery/color/named_colors.html):

```
ser.plot(
    kind="bar",
    title="Book Sales by Day",
    color="seagreen",
)
```

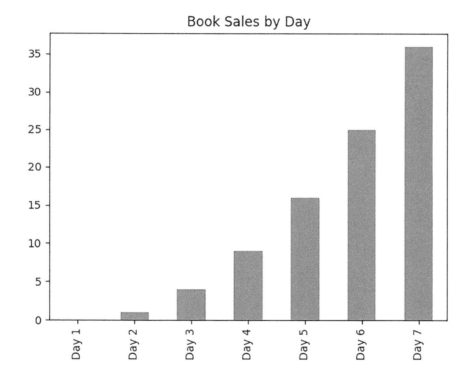

When working with a pd.DataFrame, you can pass a dictionary to pd.DataFrame.plot to control which columns should use which colors:

```python
df.plot(
    kind="bar",
    title="Book Metrics",
    color={
        "book_sales": "slateblue",
        "book_returns": "#7D5260",
    }
)
```

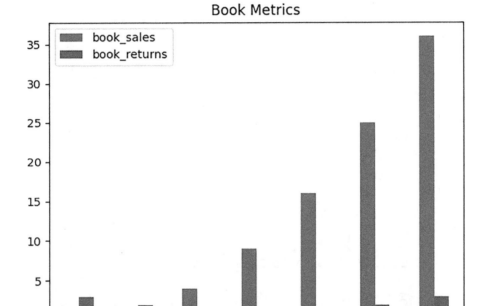

The grid= argument controls whether gridlines are shown or not:

```
ser.plot(
    kind="bar",
    title="Book Sales by Day",
    color="teal",
    grid=False,
)
```

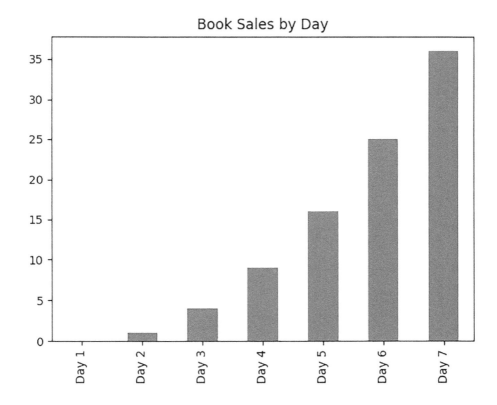

You can control how your x and y axes are labeled with the xlabel= and ylabel= arguments:

```
ser.plot(
    kind="bar",
    title="Book Sales by Day",
    color="darkgoldenrod",
    grid=False,
    xlabel="Day Number",
    ylabel="Book Sales",
)
```

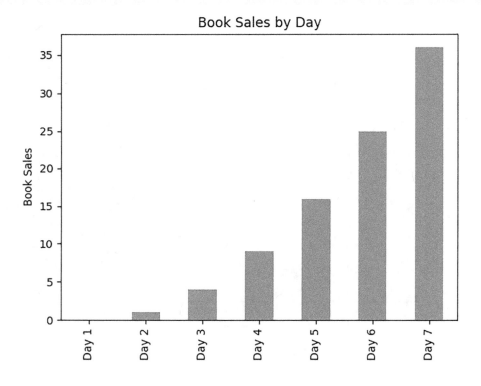

When working with a pd.DataFrame, pandas will default to placing each column's data on the same chart. However, you can easily generate separate charts with subplots=True:

```
df.plot(
    kind="bar",
    title="Book Performance",
    grid=False,
    subplots=True,
)
```

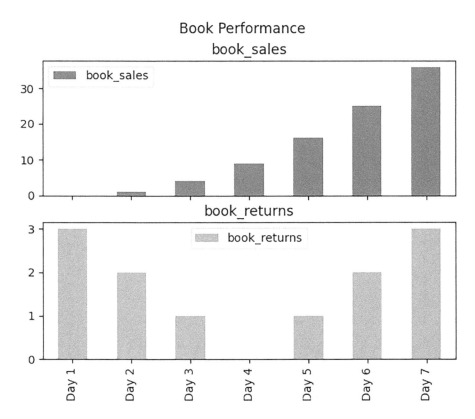

With separate charts, the legend becomes superfluous. To toggle that off, simply pass `legend=False`:

```
df.plot(
    kind="bar",
    title="Book Performance",
    grid=False,
    subplots=True,
    legend=False,
)
```

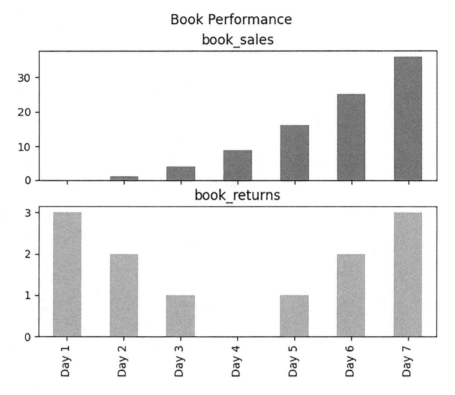

When using subplots, it is also worth noting that by default, the *x*-axis labels are shared, but the *y*-axis value ranges may differ. If you want the *y* axis to be shared, simply add `sharey=True` to your method invocation:

```
df.plot(
    kind="bar",
    title="Book Performance",
    grid=False,
    subplots=True,
    legend=False,
    sharey=True,
)
```

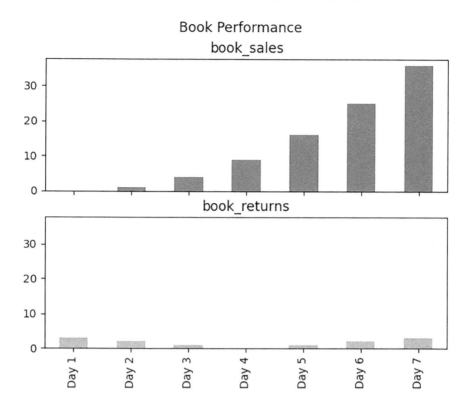

When working with pd.DataFrame.plot, the y= argument can control which columns should be visualized, which can be helpful when you don't want all of the columns to appear:

```
df.plot(
    kind="barh",
    y=["book_returns"],
    title="Book Returns",
    legend=False,
    grid=False,
    color="seagreen",
)
```

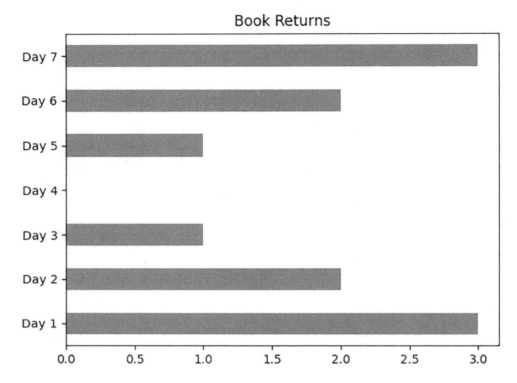

As you can see, pandas has a wealth of options to control what is being displayed and how. While pandas makes a best effort at figuring out where and how to place all of these elements on your visual, it may not always get it right. Later in this chapter, the *Further plot customization with Matplotlib* recipe will show you how to more finely control the layout of your visualization.

Plotting distributions of non-aggregated data

Visualizations can be of immense help in recognizing patterns and trends in your data. Is your data normally distributed? Does it skew left? Does it skew right? Is it multimodal? While you may be able to work out the answers to these questions, a visualization can very easily highlight these patterns for you, yielding deeper insight into your data.

In this recipe, we are going to see how easy pandas makes it to visualize the distribution of your data. Histograms are a very popular choice for plotting distributions, so we will start with them before showcasing the even more powerful **Kernel Density Estimate (KDE)** plot.

How to do it

Let's create a pd.Series using 10,000 random records that are known to follow a normal distribution. NumPy can be used to easily generate this data:

```
np.random.seed(42)
ser = pd.Series(
    np.random.default_rng().normal(size=10_000),
```

```
      dtype=pd.Float64Dtype(),
  )
  ser
```

```
0          0.049174
1         -1.577584
2         -0.597247
3           -0.0198
4          0.938997
             ...
9995      -0.141285
9996       1.363863
9997      -0.738816
9998      -0.373873
9999      -0.070183
Length: 10000, dtype: Float64
```

A histogram can be used to plot this data with the kind="hist" argument:

```
ser.plot(kind="hist")
```

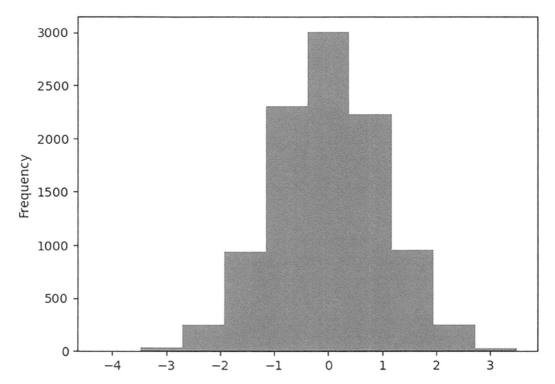

Rather than attempting to plot every single point, the histogram places our values into an automatically generated number of "bins." The range of each bin is plotted along the *X*-axis of the visualization, with the count of occurrences within each bin appearing on the y-axis of the histogram.

Since we have created the data we are visualizing, we already know that we have a normally distributed set of numbers, and the preceding histogram hints at that as well. However, we can elect to visualize a different number of bins by providing a bins= argument to pd.Series.plot, which can have a significant impact on the visualization and how it is interpreted.

To illustrate, if we were to pass bins=2, we would have so few bins that our normal distribution would not be obvious:

```
ser.plot(kind="hist", bins=2)
```

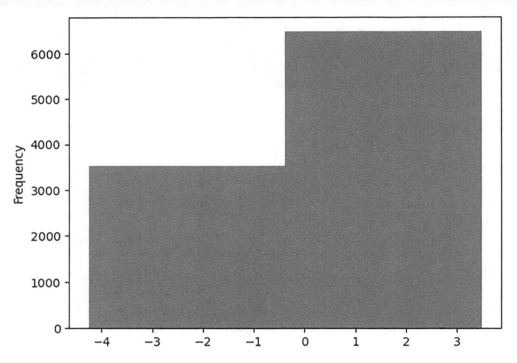

On the flip side, passing bins=100 makes it clear that we generally have a normal distribution:

```
ser.plot(kind="hist", bins=100)
```

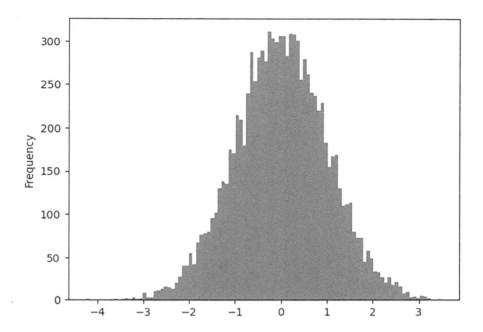

This same issue is apparent when making a histogram from a pd.DataFrame. To illustrate, let's create a pd.DataFrame with two columns, where one column is normally distributed and the other uses a triangular distribution:

```
np.random.seed(42)
df = pd.DataFrame({
    "normal": np.random.default_rng().normal(size=10_000),
    "triangular": np.random.default_rng().triangular(-2, 0, 2, size=10_000),
})
df = df.convert_dtypes(dtype_backend="numpy_nullable")

df.head()
```

```
     normal      triangular
0   -0.265525    -0.577042
1    0.327898    -0.391538
2   -1.356997    -0.110605
3    0.004558     0.71449
4    1.03956      0.676207
```

The basic plotting call to pd.DataFrame.plot will produce a chart that looks as follows:

```
df.plot(kind="hist")
```

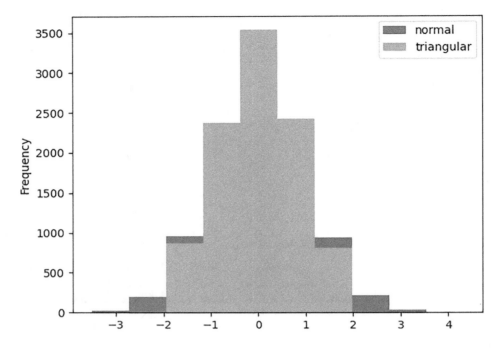

Unfortunately, the bins from one distribution overlap with the bins of the other. You can solve this by either introducing some transparency:

```
df.plot(kind="hist", alpha=0.5)
```

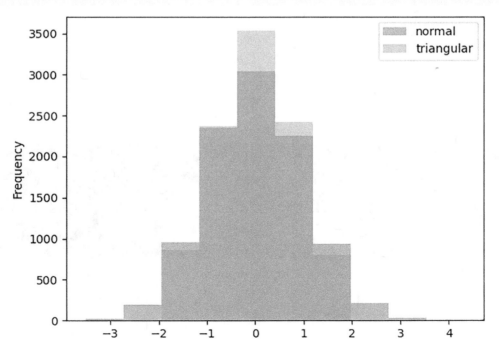

Or generating subplots:

```
df.plot(kind="hist", subplots=True)
```

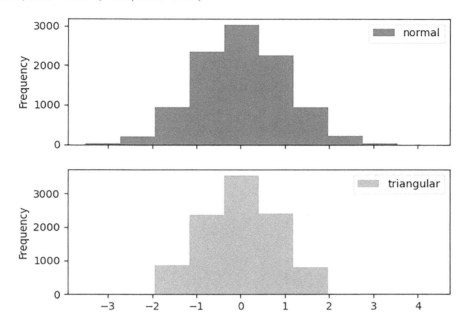

At first glance, these distributions look pretty much the same, but using more bins reveals that they are not:

```
df.plot(kind="hist", alpha=0.5, bins=100)
```

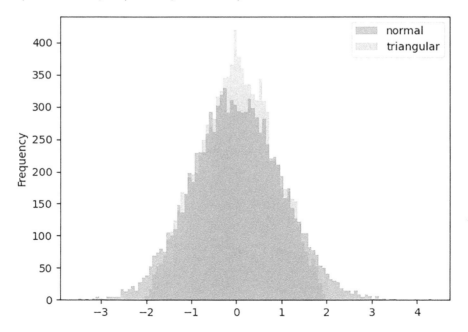

While the histogram is commonly used, the fact that the choice of binning can have an impact on the interpretation of the data is rather unfortunate; you would not want your interpretation of the data to change just from picking the "wrong" number of bins!

Fortunately, there is a similar but arguably more powerful visualization you can use that does not require you to choose any type of binning strategy, known as the **Kernel Density Estimate** (or **KDE**) plot. To use this plot, you will need to have SciPy installed:

```
python -m pip install scipy
```

After installing SciPy, you can simply pass kind="kde" to pd.Series.plot:

```
ser.plot(kind="kde")
```

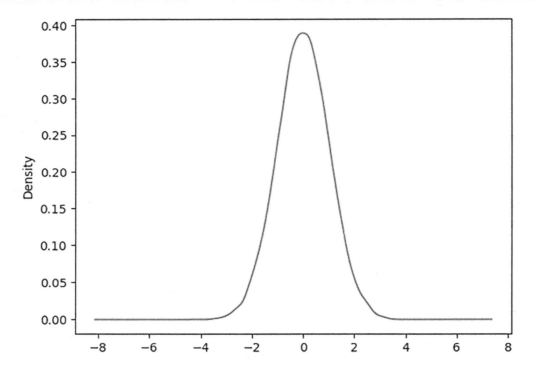

With our pd.DataFrame, the KDE plot makes it clear that we have two distinct distributions:

```
df.plot(kind="kde")
```

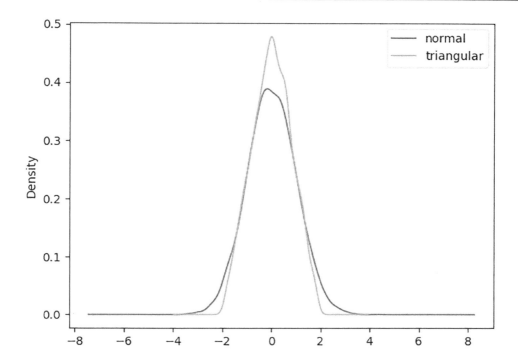

Further plot customization with Matplotlib

For very simple plots, the default layouts may suffice, but you will inevitably run into cases where you need to further tweak the generated visualization. To go beyond the out-of-the-box features in pandas, it is helpful to understand some Matplotlib terminology. In Matplotlib, the *figure* refers to the drawing area, and an *axes* or *subplot* is the region on that figure that you can draw upon. Be careful not to confuse an **axes**, which is an area for plotting data, with an **axis**, which refers to the *X*- or *Y*-axis.

How to do it

Let's start with a pd.Series of our book sales data and try to plot it three different ways on the same figure – once as a line chart, once as a bar chart, and once as a pie chart. To set up our drawing area, we will make a call to plt.subplots(nrows=1, ncols=3), essentially telling matplotlib how many rows and columns of visualizations we want in our drawing area. This will return a two-tuple containing the figure itself and a sequence of the individual Axes objects that we can plot against. We will unpack this into two variables, fig and axes, respectively.

Because we asked for one row and three columns, the length of the returned axes sequence will be three. We can pass the individual Axes we want pandas to plot on to the ax= argument of pd.DataFrame. plot. Our first attempt at drawing all of these plots should look as follows, generating a result that is, well, hideous:

```
ser = pd.Series(
    (x ** 2 for x in range(7)),
    name="book_sales",
    index=(f"Day {x + 1}" for x in range(7)),
    dtype=pd.Int64Dtype(),
)
fig, axes = plt.subplots(nrows=1, ncols=3)
ser.plot(ax=axes[0])
ser.plot(kind="bar", ax=axes[1])
ser.plot(kind="pie", ax=axes[2])
```

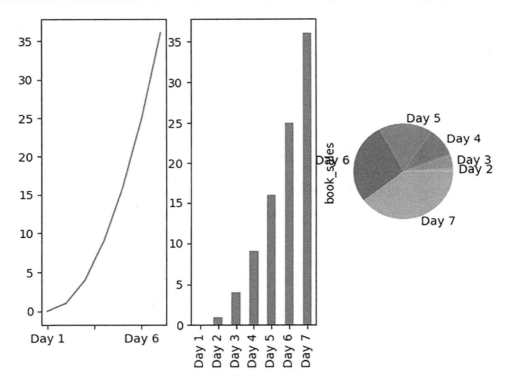

Because we did not tell it any different, Matplotlib gives us three equally sized axes objects to draw upon. However, this makes the line/bar charts above very tall and skinny, and we end up producing a ton of wasted space above and below the pie chart.

To control this more finely, we can use the Matplotlib GridSpec to create a 2x2 grid. With that, we can place our bar and line charts side by side in the first row, and then we can make the pie chart take up the entire second row:

```
from matplotlib.gridspec import GridSpec
```

```
fig = plt.figure()
gs = GridSpec(2, 2, figure=fig)
ax0 = fig.add_subplot(gs[0, 0])
ax1 = fig.add_subplot(gs[0, 1])
ax2 = fig.add_subplot(gs[1, :])
ser.plot(ax=ax0)
ser.plot(kind="bar", ax=ax1)
ser.plot(kind="pie", ax=ax2)
```

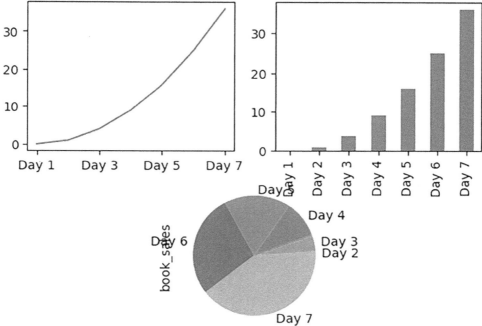

That looks a little better, but now, we still have an issue with the labels of the pie chart overlapping the X-axis labels of our bar chart. Fortunately, we can still modify each axes object individually to rotate labels, remove labels, change titles, etc.

```
from matplotlib.gridspec import GridSpec

fig = plt.figure()
fig.suptitle("Book Sales Visualized in Different Ways")
gs = GridSpec(2, 2, figure=fig, hspace=.5)
ax0 = fig.add_subplot(gs[0, 0])
ax1 = fig.add_subplot(gs[0, 1])
ax2 = fig.add_subplot(gs[1, :])
```

```
ax0 = ser.plot(ax=ax0)
ax0.set_title("Line chart")


ax1 = ser.plot(kind="bar", ax=ax1)
ax1.set_title("Bar chart")
ax1.set_xticklabels(ax1.get_xticklabels(), rotation=45)

# Remove labels from chart and show in custom legend instead
ax2 = ser.plot(kind="pie", ax=ax2, labels=None)
ax2.legend(
    ser.index,
    bbox_to_anchor=(1, -0.2, 0.5, 1),  # put legend to right of chart
    prop={"size": 6}, # set font size for legend
)
ax2.set_title("Pie Chart")
ax2.set_ylabel(None)  # remove book_sales label
```

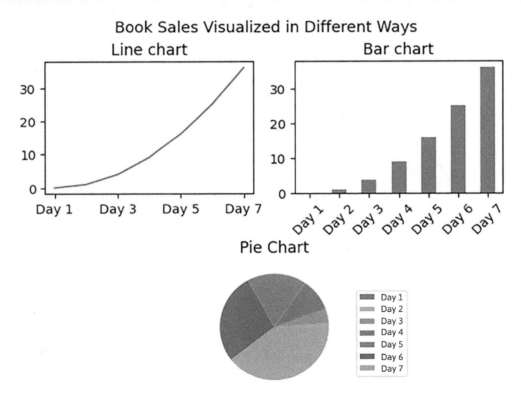

There is no limit to the amount of customization that can be done to charts via Matplotlib, and unfortunately, we cannot hope to even scratch the surface in this book. If you have a keen interest in visualizations, I highly encourage you to read the Matplotlib documentation or find a dedicated book on the topic. However, many users who just want to see their data may find the amount of customizations burdensome to handle. For those users (myself included), there are, thankfully, higher-level plotting packages like seaborn, which can produce better-looking charts with minimal extra effort. The *Using seaborn for advanced plots* recipe later in this chapter will give you an idea of just how useful that package can be.

Exploring scatter plots

Scatter plots are one of the most powerful types of visualizations that you can create. In a very compact area, a scatter plot can help you visualize the relationship between two variables, measure the scale of individual data points, and even see how these relationships and scales may vary within different categories. Being able to effectively visualize data in a scatter plot represents a significant leap in analytical capabilities when lined up against some of the more commonplace visualizations we have seen so far.

In this recipe, we will explore how we can measure all of these things at once just on one scatter plot.

How to do it

Scatter plots by definition measure the relationship of at least two variables. As such, the scatter plot can only be created with a pd.DataFrame. A pd.Series simply does not have enough variables.

With that said, let's create a sample pd.DataFrame that contains four different columns of data. Three of these columns are continuous variables, and the fourth is a color that we will eventually use to categorize different data points:

```
df = pd.DataFrame({
    "var_a": [1, 2, 3, 4, 5],
    "var_b": [1, 2, 4, 8, 16],
    "var_c": [500, 200, 600, 100, 400],
    "var_d": ["blue", "orange", "gray", "blue", "gray"],
})
df = df.convert_dtypes(dtype_backend="numpy_nullable")

df
```

	var_a	var_b	var_c	var_d
0	1	1	500	blue
1	2	2	200	orange
2	3	4	600	gray
3	4	8	100	blue
4	5	16	400	gray

Alongside kind="scatter", we will want to explicitly control what gets plotted on the *X*-axis, what gets plotted on the *Y*-axis, how big a given data point should be, and what color a given data point should appear as. These are controlled via the x=, y=, s=, and c= arguments, respectively:

```
df.plot(
    kind="scatter",
    x="var_a",
    y="var_b",
    s="var_c",
    c="var_d",
)
```

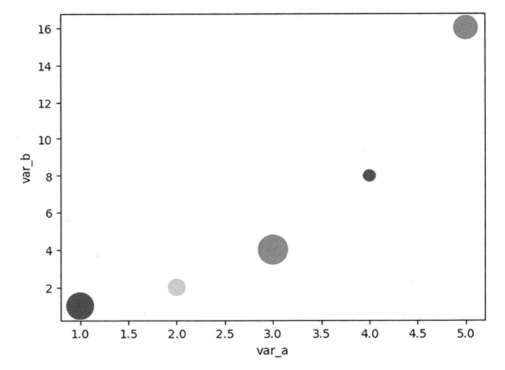

A simple scatter plot like this is not very interesting, but now that we have the basics down, let's try out a more realistic dataset. The United States Department of Energy releases annual reports (https:// www.fueleconomy.gov/feg/download.shtml) that summarize the results of detailed fuel economy testing for vehicles sold in the United States. This book includes a local copy of that dataset covering the model years 1985–2025.

For now, let's just read in a select few columns that are of interest to us, namely city08 (city miles-per-gallon), highway08 (highway miles-per-gallon), VClass (compact car, SUV, etc). fuelCost08 (annual fuel cost), and the model year of each vehicle (for a full definition of terms included with this dataset, refer to www.fueleconomy.gov):

```
df = pd.read_csv(
    "data/vehicles.csv.zip",
    dtype_backend="numpy_nullable",
    usecols=["city08", "highway08", "VClass", "fuelCost08", "year"],
)
df.head()
```

	city08	fuelCost08	highway08	VClass	year
0	19	2,450	25	Two Seaters	1985
1	9	4,700	14	Two Seaters	1985
2	23	1,900	33	Subcompact Cars	1985
3	10	4,700	12	Vans	1985
4	17	3,400	23	Compact Cars	1993

This dataset includes many different vehicle classes, so to keep our analysis focused, for now, we are just going to look at different classes of cars from 2015 onwards. Trucks, SUVs, and vans can be saved for another analysis:

```
car_classes = (
    "Subcompact Cars",
    "Compact Cars",
    "Midsize Cars",
    "Large Cars",
    "Two Seaters",
)
mask = (df["year"] >= 2015) & df["VClass"].isin(car_classes)
df = df[mask]
df.head()
```

	city08	fuelCost08	highway08	VClass	year
27058	16	3,400	23	Subcompact Cars	2015
27059	20	2,250	28	Compact Cars	2015
27060	26	1,700	37	Midsize Cars	2015
27061	28	1,600	39	Midsize Cars	2015
27062	25	1,800	35	Midsize Cars	2015

A scatter plot can be used to help us answer a question like, *What is the relationship between city miles-per-gallon and highway miles-per-gallon?* by plotting these columns on the *X*- and *Y*-axis:

```
df.plot(
    kind="scatter",
    x="city08",
    y="highway08",
)
```

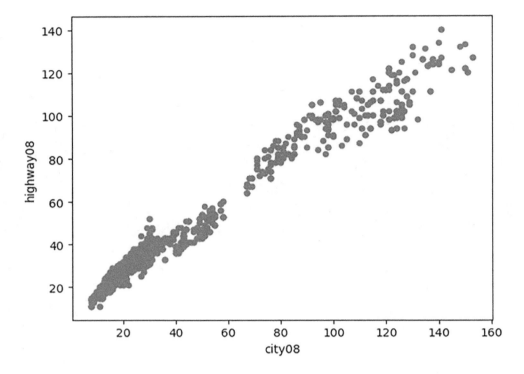

Perhaps not surprisingly, there is a strong linear trend. Chances are that the better mileage a vehicle gets on city roads, the better mileage it will get on highways.

Of course, we still see a rather large spread in values; many vehicles are clustered down in the range of 10–35 MPG, but some exceed 100. To dive in a little further, we can assign colors to each of our vehicle classes and add them to the visualization.

There are quite a few ways to do this, but one of the generally best approaches is to ensure that the value you would like to use for a color is a categorical data type:

```
classes_ser = pd.Series(car_classes, dtype=pd.StringDtype())
cat = pd.CategoricalDtype(classes_ser)
df["VClass"] = df["VClass"].astype(cat)
df.head()
```

	city08	fuelCost08	highway08	VClass	year
27058	16	3,400	23	Subcompact Cars	2015
27059	20	2,250	28	Compact Cars	2015
27060	26	1,700	37	Midsize Cars	2015
27061	28	1,600	39	Midsize Cars	2015
27062	25	1,800	35	Midsize Cars	2015

With that out of the way, you can pass the categorical column to the `c=` argument of `pd.DataFrame.plot`:

```
df.plot(
    kind="scatter",
    x="city08",
    y="highway08",
    c="VClass",
)
```

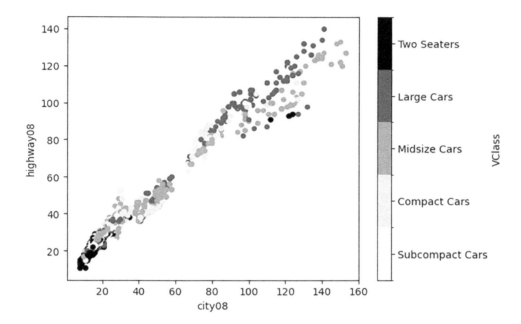

Adding a `colormap=` argument may help to visually discern data points. For a list of acceptable values for this argument, please refer to the Matplotlib documentation (`https://matplotlib.org/stable/users/explain/colors/colormaps.html`):

```
df.plot(
    kind="scatter",
    x="city08",
    y="highway08",
    c="VClass",
    colormap="Dark2",
)
```

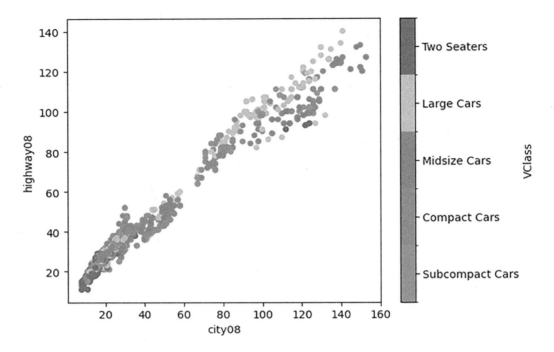

From these graphs alone, we can theorize a few things. There are not that many "Two Seaters," but when there are, they tend to do poorly on both city and highway mileage. "Midsize Cars" appear to dominate the 40–60 MPG ranges, but as you look at the vehicles that produce 100 MPG or better on both highways or cities, "Large Cars" and "Midsize Cars" both appear to be reasonably represented.

So far, we have used the *X*-axis, *Y*-axis, and color of our scatter plot to dive into data, but we can take this one step further and size each data point by fuel cost, passing `fuelCost08` as the s= argument:

```
df.plot(
    kind="scatter",
    x="city08",
    y="highway08",
    c="VClass",
    colormap="Dark2",
    s="fuelCost08",
)
```

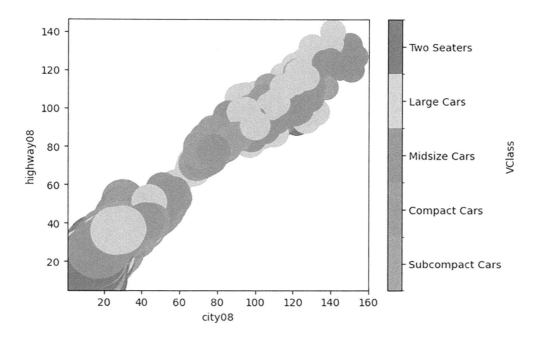

The size of the individual bubbles here is likely too large to be useful. Our fuel economy column has values that are in the range of thousands, which creates too large of a plot area to be useful. Simply scaling those values can quickly get us to a more reasonable-looking visualization; here, I have chosen to divide by 25 and introduce some transparency with the `alpha=` argument to get a more pleasing graph:

```
df.assign(
    scaled_fuel_cost=lambda x: x["fuelCost08"] / 25,
).plot(
    kind="scatter",
    x="city08",
    y="highway08",
    c="VClass",
    colormap="Dark2",
    s="scaled_fuel_cost",
    alpha=0.4,
)
```

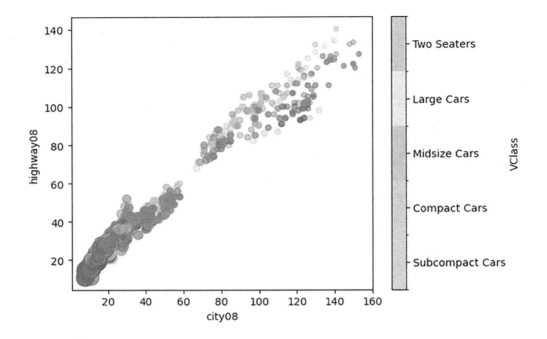

The general trend of larger circles appearing toward the origin confirms that, generally, vehicles with worse mileage have a higher annual fuel cost. You may find individual points on this scatter plot where a relatively higher mileage still has a higher fuel cost compared to other vehicles with a similar range, likely due to different fuel type requirements.

There's more...

A nice complement to the scatter plot is the scatter matrix, which generates pairwise relationships between all of the continuous columns of data within your `pd.DataFrame`. Let's see what that looks like with our vehicle data:

```
from pandas.plotting import scatter_matrix
scatter_matrix(df)
```

```
array([[<Axes: xlabel='city08', ylabel='city08'>,
        <Axes: xlabel='fuelCost08', ylabel='city08'>,
        <Axes: xlabel='highway08', ylabel='city08'>,
        <Axes: xlabel='year', ylabel='city08'>],
       [<Axes: xlabel='city08', ylabel='fuelCost08'>,
        <Axes: xlabel='fuelCost08', ylabel='fuelCost08'>,
        <Axes: xlabel='highway08', ylabel='fuelCost08'>,
        <Axes: xlabel='year', ylabel='fuelCost08'>],
       [<Axes: xlabel='city08', ylabel='highway08'>,
        <Axes: xlabel='fuelCost08', ylabel='highway08'>,
        <Axes: xlabel='highway08', ylabel='highway08'>,
        <Axes: xlabel='year', ylabel='highway08'>],
       [<Axes: xlabel='city08', ylabel='year'>,
        <Axes: xlabel='fuelCost08', ylabel='year'>,
        <Axes: xlabel='highway08', ylabel='year'>,
        <Axes: xlabel='year', ylabel='year'>]], dtype=object)
```

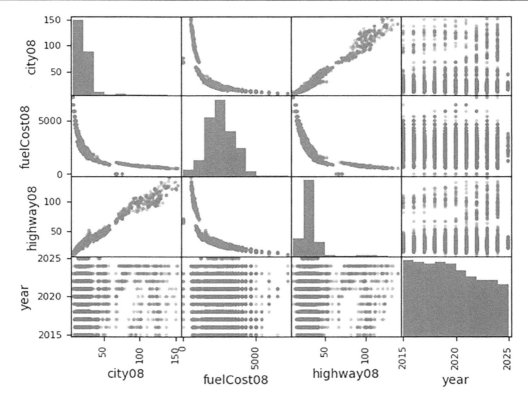

This is a lot of information in one chart, so let's start by digesting just the first column of visualizations. If you look at the bottom of the chart, the label is `city08`, which means that `city08` is the *Y*-axis for each chart in that column.

The visualization in the first row of the first column would give you the combination of `city08` on the y-axis with `city08` on the *X*-axis. Rather than a scatter plot that graphs the same column against itself, the scatter matrix shows you the distribution of `city08` values in this visual. As you can see, the majority of vehicles get less than 50 MPG in the city.

If you look one visual beneath that in the second row of the first column, you will see the relationship between fuel cost and city mileage. This would suggest that there is an exponential decrease in the amount you spend on fuel annually as you opt for cars that get better city mileage.

The visualization in the third row of the first column shows `highway08` on the *Y*-axis, which is the same visual that we displayed throughout this recipe. Once again, there is a linear relationship between city and highway mileage.

The visualization in the last row of the first column plots the year on the *Y*-axis. From this, it appears that there were more vehicles introduced in the model years 2023 and 2024 that achieved city mileage of 75 MPG and above.

Exploring categorical data

The adjective *categorical* is applied to data that, in a broad sense, is used to classify and help navigate your data, but whose values serve little to no purpose when aggregated. For example, if you were working with a dataset that contained a field called *eye color* with values of `Brown`, `Green`, `Hazel`, `Blue`, etc., you could use this field to navigate your dataset and answer questions like, *For rows where the eye color is X, what is the average pupil diameter?* However, you would not ask a question like, *What is the summation of eye color?*, as a formula like `"Hazel" + "Blue` would not make sense in this context.

By contrast, the adjective *continuous* is applied to data that you typically aggregate. With a question like, *What is the average pupil diamenter?*, the *pupil diameter* column would be considered continuous. There is value to knowing what it aggregates to (i.e., minimum, maximum, average, standard deviation, etc.), and there are a theoretically infinite amount of values that can be represented.

At times, it can be ambiguous whether your data is categorical or continuous. Using a person's *age* as an example, if you were measuring things like the *average age of subjects*, that column would be continuous, although in the context of a question like, *How many users do we have between the ages of 20–30?*, that same data becomes categorical. Ultimately, whether or not data like *age* is continuous or categorical will come down to how you use it in your analysis.

In this recipe, we are going to generate visualizations that help quickly identify the distribution of categorical data. Our next recipe, *Exploring continuous data*, will give you some ideas on how to work with continuous data instead.

How to do it

Back in the *Scatter plots* recipe, we were introduced to the `vehicles` dataset distributed by the United States Department of Energy. This dataset has a good mix of categorical and continuous data, so let's once again start by loading it into a `pd.DataFrame`:

```
df = pd.read_csv(
    "data/vehicles.csv.zip",
    dtype_backend="numpy_nullable",
)
df.head()
```

```
/tmp/ipykernel_834707/1427318601.py:1: DtypeWarning: Columns (72,74,75,77) have
mixed types. Specify dtype option on import or set low_memory=False.
  df = pd.read_csv(
    barrels08   bar-relsA08   charge120   ...   phevCity   phevHwy   phevComb
0   14.167143   0.0           0.0         ...   0          0         0
1   27.046364   0.0           0.0         ...   0          0         0
2   11.018889   0.0           0.0         ...   0          0         0
3   27.046364   0.0           0.0         ...   0          0         0
4   15.658421   0.0           0.0         ...   0          0         0
5 rows × 84 columns
```

You may have noticed that we received a warning that Columns (72,74,75,77) have mixed types. Before we jump into visualization, let's take a quick look at these columns:

```
df.iloc[:, [72, 74, 75, 77]]
```

```
        rangeA   mfrCode   c240Dscr   c240bDscr
0       <NA>     <NA>      <NA>       <NA>
1       <NA>     <NA>      <NA>       <NA>
2       <NA>     <NA>      <NA>       <NA>
3       <NA>     <NA>      <NA>       <NA>
4       <NA>     <NA>      <NA>       <NA>
...     ...      ...       ...        ...
47,518  <NA>     <NA>      <NA>       <NA>
47,519  <NA>     <NA>      <NA>       <NA>
47,520  <NA>     <NA>      <NA>       <NA>
47,521  <NA>     <NA>      <NA>       <NA>
47,522  <NA>     <NA>      <NA>       <NA>
47,523 rows × 4 columns
```

While we can see the column names, our pd.DataFrame preview does not show us any actual values, so to inspect this a bit further, we can use pd.Series.value_counts on each column.

Here is what we can see for the rangeA column:

```
df["rangeA"].value_counts()
```

```
rangeA
290              74
270              58
280              56
310              41
277              38
                 ..
240/290/290      1
395              1
258              1
256              1
230/350          1
Name: count, Length: 264, dtype: int64
```

The values here are... interesting. Without yet knowing what we are looking at, the column name rangeA and most of the values suggest there is some value to treating this as continuous. By doing so, we could answer questions like, *What is the average rangeA of vehicles that...?*, but the presence of values like 240/290/290 and 230/350 that we see will prevent us from being able to do that. For now, we are just going to treat this data as a string.

To bring us full circle on the warning issued by pd.read_csv, pandas tries to infer the data type while reading the CSV file. If much of the data at the beginning of the file shows one type but later in the file you see another type, pandas will intentionally throw this warning so that you are aware of any potential issues with your data. For this column, we can use pd.Series.str.isnumeric alongside pd.Series.idxmax to quickly determine the first row where a non-integral value was encountered in the CSV file:

```
df["rangeA"].str.isnumeric().idxmax()
```

```
7116
```

If you were to inspect the other columns that pd.read_csv warned about, you would not see a mix of integral and string data, but you would see that much of the data at the beginning of the file is missing, which makes it difficult for pandas to infer the data type:

```
df.iloc[:, [74, 75, 77]].pipe(pd.isna).idxmin()
```

```
mfrCode       23147
c240Dscr      25661
c240bDscr     25661
dtype: int64
```

Of course, the best solution here would have been to avoid the use of CSV files in the first place, opting instead for a data storage format that maintains type metadata, like Apache Parquet. However, we have no control over how this data is generated, so the best we can do for now is explicitly tell pd.read_csv to treat all of these columns as strings and suppress any warnings:

```
df = pd.read_csv(
    "data/vehicles.csv.zip",
    dtype_backend="numpy_nullable",
    dtype={
        "rangeA": pd.StringDtype(),
        "mfrCode": pd.StringDtype(),
        "c240Dscr": pd.StringDtype(),
        "c240bDscr": pd.StringDtype()
    }
)
df.head()
```

	barrels08	bar-relsA08	charge120	...	phevCity	phevHwy	phevComb
0	14.167143	0.0	0.0	...	0	0	0
1	27.046364	0.0	0.0	...	0	0	0
2	11.018889	0.0	0.0	...	0	0	0
3	27.046364	0.0	0.0	...	0	0	0
4	15.658421	0.0	0.0	...	0	0	0

5 rows × 84 columns

Now that we have loaded the data cleanly, let's try and identify columns that are categorical in nature. Since we know nothing about this dataset, we can make the directionally correct assumption that all columns read in as strings by `pd.read_csv` are categorical in nature:

```
df.select_dtypes(include=["string"]).columns
```

```
Index(['drive', 'eng_dscr', 'fuelType', 'fuelType1', 'make', 'model',
       'mpgData', 'trany', 'VClass', 'baseModel', 'guzzler', 'trans_dscr',
       'tCharger', 'sCharger', 'atvType', 'fuelType2', 'rangeA', 'evMotor',
       'mfrCode', 'c240Dscr', 'c240bDscr', 'createdOn', 'modifiedOn',
       'startStop'],
      dtype='object')
```

We could loop over all of these columns and call `pd.Series.value_counts` to understand what each column contains, but a more effective way to explore this data would be to first understand how many unique values are in each column with `pd.Series.nunique`, ordering from low to high. A lower number indicates a *low cardinality* (i.e., the number of unique values compared to the value count of the `pd.DataFrame` is relatively low). Fields with a higher number would inversely be considered to have a *high cardinality*:

```
df.select_dtypes(include=["string"]).nunique().sort_values()
```

```
    sCharger        1
    tCharger        1
    startStop       2
    mpgData         2
    guzzler         3
    fuelType2       4
    c240Dscr        5
    c240bDscr       7
    drive           7
    fuelType1       7
    atvType         9
    fuelType       15
    VClass         34
    trany          40
    trans_dscr     52
    mfrCode        56
    make          144
    rangeA        245
    modifiedOn    298
    evMotor       400
    createdOn     455
    eng_dscr      608
    baseModel    1451
    model        5064
    dtype: int64
```

For an easy visualization, we are just going to take the nine columns with the lowest cardinality. This is by no means an absolute rule for choosing what to visualize or not – ultimately, that decision is up to you. For our particular dataset, the nine columns with the lowest cardinality have up to seven unique values, which can be reasonably plotted on the *X*-axis of bar charts to help visualize value distribution.

Building on what we learned back in the *Further plot customization with Matplotlib* recipe in this chapter, we can use plt.subplots to create a simple 3x3 grid and, with that, plot each visual to its own space in the grid:

```
low_card = df.select_dtypes(include=["string"]).nunique().sort_values().
iloc[:9].index
fig, axes = plt.subplots(nrows=3, ncols=3)

for index, column in enumerate(low_card):
    row = index % 3
    col = index // 3
```

```
    ax = axes[row][col]
    df[column].value_counts().plot(kind="bar", ax=ax)

plt.tight_layout()
```

```
/tmp/ipykernel_834707/4000549653.py:10: UserWarning: Tight layout not applied.
tight_layout cannot make axes height small enough to accommodate all axes
decorations.
 plt.tight_layout()
```

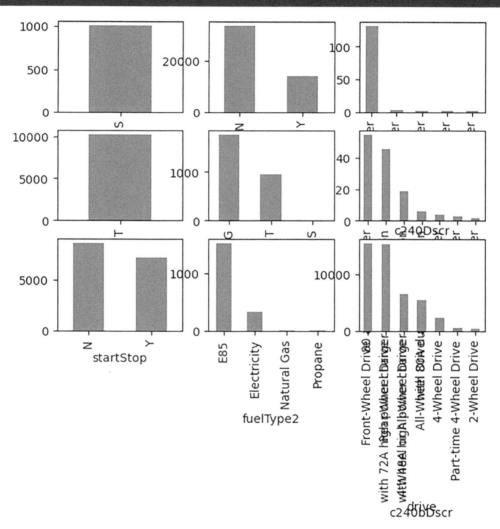

That chart is... very difficult to read. Many of the *X*-axis labels overrun the chart area, given their length. One way to fix this would be to assign shorter labels to our row index values, using a combination of pd.Index.str[] along with pd.Index.set_axis to use those values in a new pd.Index. We can also use Matplotlib to rotate and resize our *X*-axis labels:

```
low_card = df.select_dtypes(include=["string"]).nunique().sort_values().
iloc[:9].index
fig, axes = plt.subplots(nrows=3, ncols=3)

for index, column in enumerate(low_card):
    row = index % 3
    col = index // 3
    ax = axes[row][col]
    counts = df[column].value_counts()
    counts.set_axis(counts.index.str[:8]).plot(kind="bar", ax=ax)
    ax.set_xticklabels(ax.get_xticklabels(), rotation=45, fontsize=6)

plt.tight_layout()
```

From this visualization, we can more easily understand our dataset at a high level. The mpgData column appears to be N at a much higher frequency than Y. For the guzzler column, we see roughly twice as many G values as T. For the c240Dscr column, we can see that the vast majority of entries are standard, although overall, there are only slightly more than 100 rows in our entire dataset that even bother to assign this value, so we may decide that there aren't enough measurements to reliably use it.

Exploring continuous data

In the *Exploring categorical data* recipe, we provided a definition for *categorical* and *continuous* data, while exploring only the former. The same vehicles dataset we used in that recipe has a good mix of both types of data (most datasets will), so we will reuse that same dataset but shift our focus to continuous data for this recipe.

Before going through this recipe, I advise you to get familiar with the techniques shown in the *Plotting distributions of non-aggregated data* recipe first. The actual plotting calls made will be the same, but this recipe will apply them to more of a "real-world" dataset instead of artificially created data.

How to do it

Let's start by loading the vehicles dataset:

```
df = pd.read_csv(
    "data/vehicles.csv.zip",
    dtype_backend="numpy_nullable",
    dtype={
        "rangeA": pd.StringDtype(),
        "mfrCode": pd.StringDtype(),
        "c240Dscr": pd.StringDtype(),
        "c240bDscr": pd.StringDtype()
    }
)
df.head()
```

```
     barrels08  bar-relsA08  charge120  ...  phevCity  phevHwy  phevComb
0    14.167143  0.0          0.0        ...  0         0        0
1    27.046364  0.0          0.0        ...  0         0        0
2    11.018889  0.0          0.0        ...  0         0        0
3    27.046364  0.0          0.0        ...  0         0        0
4    15.658421  0.0          0.0        ...  0         0        0
5 rows × 84 columns
```

In the previous recipe, we used pd.DataFrame.select_dtypes with an include= argument that kept only string columns, which we used as a proxy for categorical data. By passing that same argument to exclude= instead, we can get a reasonable overview of the continuous columns:

```
df.select_dtypes(exclude=["string"]).columns
```

```
Index(['barrels08', 'barrelsA08', 'charge120', 'charge240', 'city08',
       'city08U', 'cityA08', 'cityA08U', 'cityCD', 'cityE', 'cityUF', 'co2',
       'co2A', 'co2TailpipeAGpm', 'co2TailpipeGpm', 'comb08', 'comb08U',
       'combA08', 'combA08U', 'combE', 'combinedCD', 'combinedUF', 'cylinders',
       'displ', 'engId', 'feScore', 'fuelCost08', 'fuelCostA08', 'ghgScore',
       'ghgScoreA', 'highway08', 'highway08U', 'highwayA08', 'highwayA08U',
       'highwayCD', 'highwayE', 'highwayUF', 'hlv', 'hpv', 'id', 'lv2', 'lv4',
       'phevBlended', 'pv2', 'pv4', 'range', 'rangeCity', 'rangeCityA',
       'rangeHwy', 'rangeHwyA', 'UCity', 'UCityA', 'UHighway', 'UHighwayA',
       'year', 'youSaveSpend', 'charge240b', 'phevCity', 'phevHwy',
       'phevComb'],
      dtype='object')
```

Using pd.Series.nunique does not make as much sense with continuous data, as values may take on a theoretically infinite amount of values. Instead, to identify good plotting candidates, we may just want to understand which columns have a sufficient amount of non-missing data by using pd.isna:

```python
df.select_dtypes(
    exclude=["string"]
).pipe(pd.isna).sum().sort_values(ascending=False).head()
```

```
cylinders      801
displ          799
barrels08        0
pv4              0
highwayA08U      0
dtype: int64
```

Generally, most of our continuous data is complete, but let's take a look at cylinders to see what the missing values are:

```python
df.loc[df["cylinders"].isna(), ["make", "model"]].value_counts()
```

```
make      model
Fiat      500e                           8
smart     fortwo electric drive coupe    7
Toyota    RAV4 EV                        7
Nissan    Leaf                           7
Ford      Focus Electric                 7
                                        ..
Polestar  2 Single Motor (19 Inch Wheels)  1
Ford      Mustang Mach-E RWD LFP         1
Polestar  2 Dual Motor Performance Pack  1
          2 Dual Motor Perf Pack         1
Acura     ZDX AWD                        1
Name: count, Length: 450, dtype: int64
```

These appear to be electric vehicles, so we could reasonably choose to fill these missing values with a 0 instead:

```
df["cylinders"] = df["cylinders"].fillna(0)
```

We see the same pattern with the `displ` column:

```
df.loc[df["displ"].isna(), ["make", "model"]].value_counts()
```

```
make     model
Fiat     500e                                8
smart    fortwo electric drive coupe         7
Toyota   RAV4 EV                             7
Nissan   Leaf                                7
Ford     Focus Electric                      7
                                            ..
Porsche  Taycan 4S Performance Battery Plus  1
         Taycan GTS ST                       1
Fisker   Ocean Extreme One                   1
Fiat     500e All Season                     1
Acura    ZDX AWD                             1
Name: count, Length: 449, dtype: int64
```

Whether or not we should fill this data with 0 is up for debate. In the case of `cylinder`, filling missing values with 0 made sense because our data was actually categorical (i.e., there are only so many `cylinder` values that can appear, and you cannot simply aggregate those values). If you have one vehicle with 2 cylinders and another with 3, it would not make sense to say, "The average number of cylinders is 2.5" because a vehicle may not have 2.5 cylinders.

However, with a column like `displacement`, it may make more sense to measure something like the "average displacement." In such a case, providing many 0 values to an average will skew it downwards, whereas missing values would be ignored. There are also many more unique values than what we see with `cylinders`:

```
df["displ"].nunique()
```

```
66
```

Ultimately, filling missing values in this field is a judgment call; for our analysis, we will leave them as missing.

Now that we have validated the missing values in our dataset and feel comfortable with our completeness, it is time to start exploring individual fields in more detail. When exploring continuous data, a histogram is often the first visualization that users reach for. Let's see what that looks like with our `city08` column:

```
df["city08"].plot(kind="hist")
```

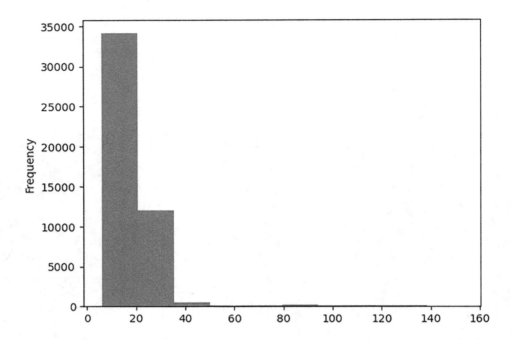

The plot looks very skewed, so we will increase the number of bins in the histogram to see if the skew is hiding behaviors (as skew makes bins wider):

```
df["city08"].plot(kind="hist", bins=30)
```

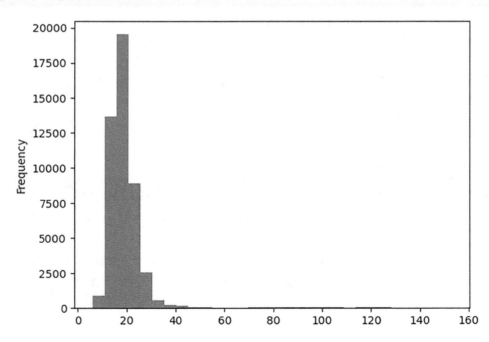

As we discussed back in the *Plotting distributions of non-aggregated data* recipe, you can forgo having to find the optimal number of bins if you have SciPy installed. With SciPy, a KDE plot will give you an even better view of the distribution.

Knowing that, and building from what we saw back in the *Further plot customization with Matplotlib* recipe, we can use `plt.subplots` to visualize the KDE plots for multiple variables at once, like city and highway mileage:

```
fig, axes = plt.subplots(nrows=2, ncols=1)
axes[0].set_xlim(0, 40)
axes[1].set_xlim(0, 40)
axes[0].set_ylabel("city")
axes[1].set_ylabel("highway")

df["city08"].plot(kind="kde", ax=axes[0])
df["highway08"].plot(kind="kde", ax=axes[1])
```

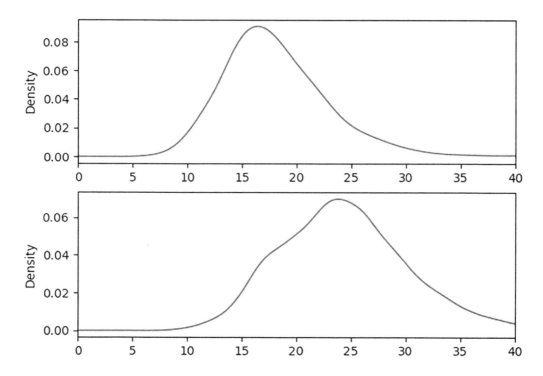

As you can see, the city mileage tends to skew slightly left, with the peak of the distribution occurring around 16 or 17 mpg. Highway mileage peaks closer to 23 or 24, with more values than you would expect for a perfectly normal distribution appearing around 17 or 18 mpg.

Using seaborn for advanced plots

The seaborn library is a popular Python library for creating visualizations. It does not do any drawing itself, instead deferring the heavy lifting to Matplotlib. However, for users working with a `pd.DataFrame`, seaborn offers beautiful visualizations out of the box and an API that abstracts a lot of things you would have to do when working more directly with Matplotlib.

Rather than using `pd.Series.plot` and `pd.DataFrame.plot`, we will use seaborn's own API. All examples in this section assume the following code to import seaborn and use its default theme:

```python
import seaborn as sns
sns.set_theme()
sns.set_style("white")
```

How to do it

Let's create a small `pd.DataFrame` that shows how many stars two GitHub projects have received over time:

```python
df = pd.DataFrame([
    ["Q1-2024", "project_a", 1],
    ["Q1-2024", "project_b", 1],
    ["Q2-2024", "project_a", 2],
    ["Q2-2024", "project_b", 2],
    ["Q3-2024", "project_a", 4],
    ["Q3-2024", "project_b", 3],
    ["Q4-2024", "project_a", 8],
    ["Q4-2024", "project_b", 4],
    ["Q5-2025", "project_a", 16],
    ["Q5-2025", "project_b", 5],
], columns=["quarter", "project", "github_stars"])
df = df.convert_dtypes(dtype_backend="numpy_nullable")

df
```

```
   quarter    project     github_stars
0  Q1-2024    project_a   1
1  Q1-2024    project_b   1
2  Q2-2024    project_a   2
3  Q2-2024    project_b   2
4  Q3-2024    project_a   4
5  Q3-2024    project_b   3
6  Q4-2024    project_a   8
7  Q4-2024    project_b   4
8  Q5-2025    project_a   16
9  Q5-2025    project_b   5
```

This simple data makes for a good bar chart, which we can produce with `sns.barplot`. Note the difference in the call signature when using seaborn's API – with seaborn, you will provide the `pd.DataFrame` as an argument and explicitly choose the x, y, and hue arguments. You will also notice that the seaborn theme uses a different color theme than Matplotlib, which you may find more visually appealing:

```
sns.barplot(df, x="quarter", y="github_stars", hue="project")
```

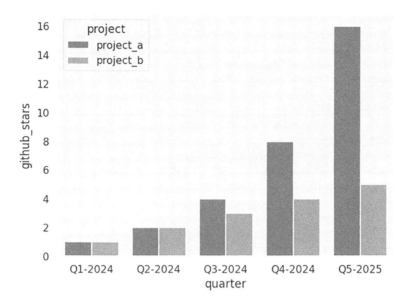

`sns.lineplot` can be used to produce this same visualization as a line chart:

```
sns.lineplot(df, x="quarter", y="github_stars", hue="project")
```

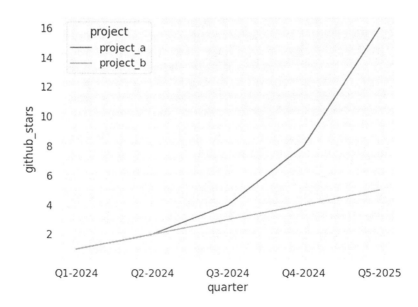

An important thing to note when using seaborn is that you will provide your data in *long form* instead of wide form. To illustrate the difference, look closely at the raw pd.DataFrame we just plotted:

```
df
```

```
     quarter    project    github_stars
0    Q1-2024    project_a   1
1    Q1-2024    project_b   1
2    Q2-2024    project_a   2
3    Q2-2024    project_b   2
4    Q3-2024    project_a   4
5    Q3-2024    project_b   3
6    Q4-2024    project_a   8
7    Q4-2024    project_b   4
8    Q5-2025    project_a   16
9    Q5-2025    project_b   5
```

If we wanted to make the equivalent line and bar charts with pandas, we would have had to structure our data differently before calling pd.DataFrame.plot:

```
df = pd.DataFrame({
    "project_a": [1, 2, 4, 8, 16],
    "project_b": [1, 2, 3, 4, 5],
}, index=["Q1-2024", "Q2-2024", "Q3-2024", "Q4-2024", "Q5-2024"])
df = df.convert_dtypes(dtype_backend="numpy_nullable")
```

```
df
```

```
           project_a    project_b
Q1-2024    1            1
Q2-2024    2            2
Q3-2024    4            3
Q4-2024    8            4
Q5-2024    16           5
```

While the default styling that seaborn provides is helpful to make nice-looking charts, there are way more powerful visualizations that the library can help you build, with no equivalent when using pandas directly.

To see these types of charts in action, let's once again work with the movie dataset we explored back in *Chapter 5, Algorithms and How to Apply Them*:

```
df = pd.read_csv(
    "data/movie.csv",
    usecols=["movie_title", "title_year", "imdb_score", "content_rating"],
```

```
    dtype_backend="numpy_nullable",
)
df.head()
```

```
   movie_title  content_rating  title_year  imdb_score
0  Avatar           PG-13           2009.0      7.9
1  Pirates of the Caribbean: At World's End    PG-13    2007.0   7.1
2  Spectre          PG-13           2015.0      6.8
3  The Dark Knight Rises    PG-13   2012.0   8.5
4  Star Wars: Episode VII - The Force Awakens   <NA>   <NA>   7.1
```

We need to do some data cleansing before we jump into this dataset. For starters, it looks like `title_year` is being read as a floating-point value. An integral value would have been much more appropriate, so we are going to reread our data while passing that explicitly to the `dtype=` argument:

```
df = pd.read_csv(
    "data/movie.csv",
    usecols=["movie_title", "title_year", "imdb_score", "content_rating"],
    dtype_backend="numpy_nullable",
    dtype={"title_year": pd.Int16Dtype()},
)
df.head()
```

```
   movie_title  content_rating  title_year  imdb_score
0  Avatar           PG-13           2009        7.9
1  Pirates of the Caribbean: At World's End    PG-13    2007   7.1
2  Spectre          PG-13           2015        6.8
3  The Dark Knight Rises    PG-13   2012   8.5
4  Star Wars: Episode VII - The Force Awakens   <NA>   <NA>   7.1
```

With that out of the way, let's see when the oldest movie in our dataset was released:

```
df["title_year"].min()
```

```
1916
```

And compare that to the last movie:

```
df["title_year"].max()
```

```
2016
```

As we think ahead toward visualizing this data, we may not always be so detailed as caring about the exact year that a movie was released. Instead, we could place each movie into a `decade` bucket by using the `pd.cut` function we covered back in *Chapter 5, Algorithms and How to Apply Them*, providing a range that will start before and extend after the first and last titles in our dataset were released:

```
df = df.assign(
    title_decade=lambda x: pd.cut(x["title_year"],
                          bins=range(1910, 2021, 10)))

df.head()
```

```
   movie_title     content_rating   title_year   imdb_score   title_decade
0  Avatar          PG-13            2009         7.9          (2000.0, 2010.0]
1  Pirates of the Caribbean: At World's End  PG-13   2007   7.1   (2000.0,
2010.0]
2  Spectre         PG-13            2015         6.8          (2010.0, 2020.0]
3  The Dark Knight Rises   PG-13    2012         8.5          (2010.0, 2020.0]
4  Star Wars: Epi-sode VII - The Force Awakens   <NA>   <NA>   7.1   NaN
```

If we wanted to understand how the distribution of movie ratings has changed over the decades, a boxplot would be a great first step toward visualizing those trends. Seaborn exposes an sns.boxplot method that makes this trivial to draw:

```
sns.boxplot(
    data=df,
    x="imdb_score",
    y="title_decade",
)
```

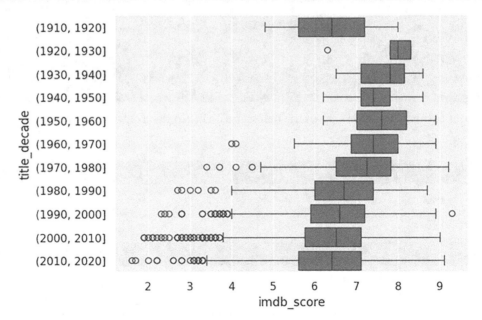

If you look at the median movie rating (i.e., the black vertical line toward the middle of each box), you can see that movie ratings have generally trended downward over time. At the same time, the lines extending from each box (which represent the lowest and highest quartiles of ratings) appear to have a wider spread over time, which may suggest that the worst movies each decade are getting worse, while the best movies may be getting better, at least since the 1980s.

While the boxplot chart provides a decent high-level view into the distribution of data by decade, there are other plots that seaborn offers that may be even more insightful. One example is the violin plot, which is essentially a KDE plot (covered back in the *Plotting distributions of non-aggregated data* recipe) overlaid on top of a boxplot. Seaborn allows this via the `sns.violinplot` function:

```
sns.violinplot(
    data=df,
    x="imdb_score",
    y="title_decade",
)
```

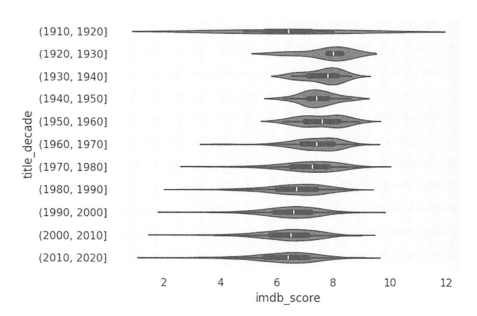

Many of the decades show a distribution with a single mode, but if you look closely at the 1950s, you will notice that the KDE plot has two peaks (one around a score of 7 and the other peak slightly north of 8). The 1960s exhibit a similar phenomenon, although the peak around a score of 7 is slightly less pronounced. For both of these decades, the KDE overlay suggests that a relatively high volume of reviews are distributed toward the 25th and 75th percentiles for ratings, whereas other decades tend to regress more toward the median.

However, the violin plot still makes it difficult to discern how many ratings there were per decade. While the distribution in each decade is, of course, important, the volume may tell another story. Perhaps movie ratings are higher for older decades because of survivorship bias, with only the movies that are deemed good for those decades actually getting reviewed, or perhaps newer decades have valued quality over quantity.

Whatever the root cause may be, seaborn can at least help us visually confirm the volume alongside our distribution through the use of a swarm plot, which takes the KDE portion of the violin plot and scales it vertically, depending on the volume of records:

```
sns.swarmplot(
    data=df,
    x="imdb_score",
    y="title_decade",
    size=.25,
)
```

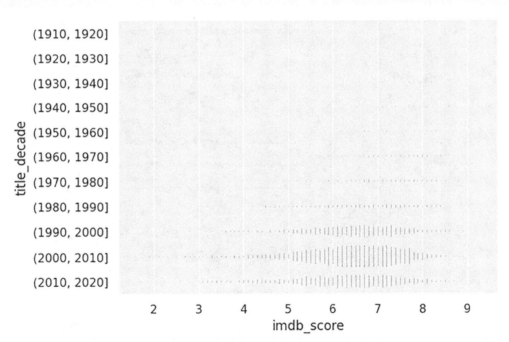

As you can see in the visual, much of the volume of reviews happened on reviews from the 1990s or later, with most of the reviews coming from the years 2000–2010 (remember that our dataset only contains movies up until 2016). Decades before 1990 have a relatively small amount of reviews, to the point of making them almost indiscernible on the graph.

With a swarm plot, you can go even further into this data by adding more dimensions to the visual. So far, we have already discovered that movie ratings have trended downward over time, whether it is due to survivorship bias with ratings or a focus on quantity over quality. But what if we wanted to know more about different types of movies? Are PG-13 movies faring better than R-rated?

By controlling the color of each point on a swarm plot, you can add that extra dimension to your visuals. To see this in action, let's drill into just a few years of data, as our current plots are getting tough to read. We can also just look at movies with ratings, as unrated entries or TV shows are not something we care to drill into. As a final data cleansing step, we are going to convert our title_year column to a categorical data type so that the plotting library knows that years like 2013, 2014, 2015, etc. should be treated as discrete values, rather than as a continuous range of values from 2013 to 2015:

```python
ratings_of_interest = {"G", "PG", "PG-13", "R"}
mask = (
    (df["title_year"] >= 2013)
    & (df["title_year"] <= 2015)
    & (df["content_rating"].isin(ratings_of_interest))
)
data = df[mask].assign(
    title_year=lambda x: x["title_year"].astype(pd.CategoricalDtype())
)
data.head()
```

	movie_title	content_rating	title_year	imdb_score	title_decade
2	Spectre	PG-13	2015	6.8	(2010, 2020]
8	Avengers: Age of Ultron	PG-13	2015	7.5	(2010, 2020]
14	The Lone Ranger	PG-13	2013	6.5	(2010, 2020]
15	Man of Steel	PG-13	2013	7.2	(2010, 2020]
20	The Hobbit: The Battle of the Five Ar-mies	PG-13	2014	7.5	(2010, 2020]

With the data cleansing out of the way, we can go ahead and add the content_rating to our chart and have seaborn assign each a unique color, via the hue= argument:

```python
sns.swarmplot(
    data=data,
    x="imdb_score",
    y="title_year",
    hue="content_rating",
)
```

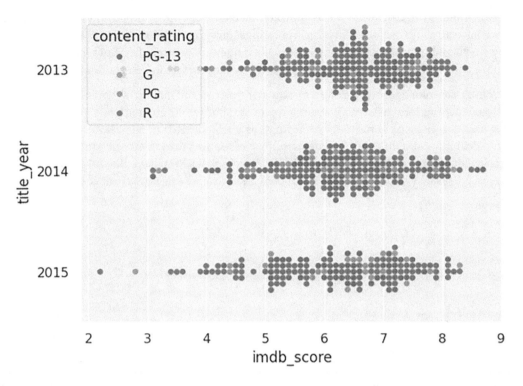

Adding colors adds another dimension of information to our chart, although an alternative approach to use a separate chart for each content_rating might make this even more readable.

To achieve that, we are going to use the sns.catplot function with some extra arguments. The first argument of note is kind=, through which we will tell seaborn to draw "swarm" plots for us. The col= argument dictates the field used to generate individual charts, and the col_wrap= argument tells us how many charts can be put together in a row, assuming a grid-like layout for our charts:

```
sns.catplot(
    kind="swarm",
    data=data,
    x="imdb_score",
    y="title_year",
    col="content_rating",
    col_wrap=2,
)
```

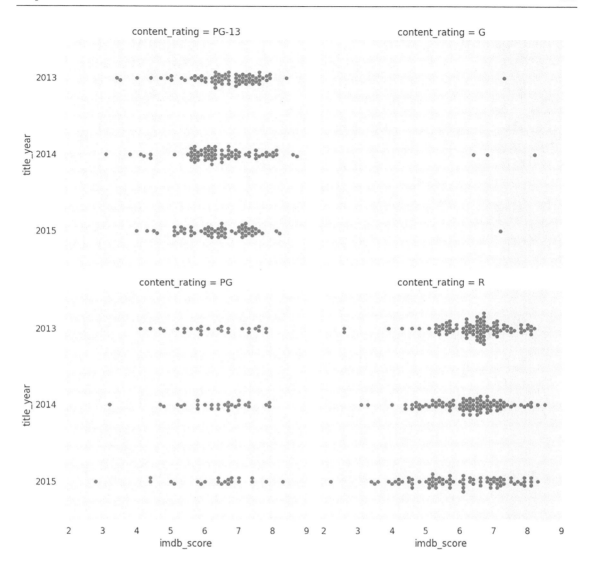

This visualization would appear to suggest that 2013 was a good year for movies, at least relative to 2014 and 2015. Within the PG-13 content rating, it appears that there were relatively more movies falling in the 7–8 range than any other year. For R-rated movies, it appears that the vast majority of movies were given a rating of 5 and above in 2013, with more movies falling below that line as the years went on.

Unlock this book's exclusive benefits now

This book comes with additional benefits designed to elevate your learning experience.

Note: Have your purchase invoice ready before you begin.   `https://www.packtpub.com/unlock/9781836205876`

7

Reshaping DataFrames

Working with data is hard. Rarely, if ever, can you just collect data and have it immediately yield insights. Often, significant time and effort must be put into cleansing, transforming, and *reshaping* your data to get it into a format that is usable, digestible, and/or understandable.

Is your source data a collection of CSV files, where each file represents a different day? Proper use of pd.concat will help you take those files and combine them into one with ease.

Does the relational database you use as a source store data in a normalized form, while the target columnar database would prefer to ingest data all in one table? pd.merge can help you combine your data together.

What if your boss asks you to take millions of rows of data and, from that, produce a nice summary report that anyone in the business can understand? pd.pivot_table is the right tool for the job here, allowing you to quickly and easily summarize your data.

Ultimately, the reasons why you need to reshape your data come from different places. Whether your requirements are driven by systems or people, pandas can help you manipulate data as needed.

Throughout this chapter, we will walk you through the functions and methods that pandas offers to reshape your data. Equipped with the proper knowledge and some creativity, reshaping with pandas can be one of the most fun and rewarding parts of your analytical process.

We are going to cover the following recipes in this chapter:

- Concatenating pd.DataFrame objects
- Merging DataFrames with pd.merge
- Joining DataFrames with pd.DataFrame.join
- Reshaping with pd.DataFrame.stack and pd.DataFrame.unstack
- Reshaping with pd.DataFrame.melt
- Reshaping with pd.wide_to_long
- Reshaping with pd.DataFrame.pivot and pd.pivot_table
- Reshaping with pd.DataFrame.explode
- Transposing with pd.DataFrame.T

Concatenating pd.DataFrame objects

The term *concatenation* in pandas refers to the process of taking two or more pd.DataFrame objects and stacking them in some manner. Most commonly, users in pandas perform what we would consider to be *vertical* concatenation, which places the pd.DataFrame objects on top of one another:

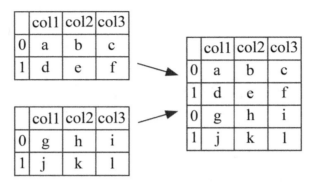

Figure 7.1: Vertical concatenation of two pd.DataFrame objects

However, pandas also has the flexibility to take your pd.DataFrame objects and stack them side by side, through a process called *horizontal* concatenation:

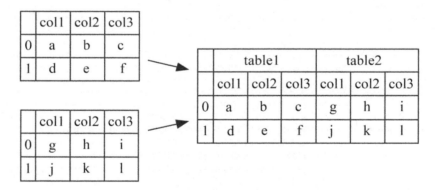

Figure 7.2: Vertical concatenation of two pd.DataFrame objects

These figures may provide you with a good grasp of what concatenation is all about, but there are some potential issues to consider. What should happen if we try to concatenate vertically, but our column labels are not the same across all of the objects? On the flip side, what should happen if we try to concatenate horizontally, and not all of the row labels are the same?

Regardless of the direction along which you would like to concatenate, and regardless of how your labels may or may not align, concatenation in pandas is controlled entirely through the pd.concat function. This recipe will walk you through the basics of pd.concat, while showing you how you can control its behavior when you aren't always working with like-labeled pd.DataFrame objects.

How to do it

Let's imagine we have collected data about the stock performance of various companies across two different quarters. To best showcase how concatenation works, we have intentionally made it so that the two pd.DataFrame objects cover different time periods, show different companies, and even contain different columns:

```
df_q1 = pd.DataFrame([
    ["AAPL", 100., 50., 75.],
    ["MSFT", 80., 42., 62.],
    ["AMZN", 60., 100., 120.],
], columns=["ticker", "shares", "low", "high"])
df_q1 = df_q1.convert_dtypes(dtype_backend="numpy_nullable")
```

```
df_q1
```

	ticker	shares	low	high
0	AAPL	100	50	75
1	MSFT	80	42	62
2	AMZN	60	100	120

```
df_q2 = pd.DataFrame([
    ["AAPL", 80., 70., 80., 77.],
    ["MSFT", 90., 50., 60., 55.],
    ["IBM", 100., 60., 70., 64.],
    ["GE", 42., 30., 50., 44.],
], columns=["ticker", "shares", "low", "high", "close"])
df_q2 = df_q2.convert_dtypes(dtype_backend="numpy_nullable")
```

```
df_q2
```

	ticker	shares	low	high	close
0	AAPL	80	70	80	77
1	MSFT	90	50	60	55
2	IBM	100	60	70	64
3	GE	42	30	50	44

The most basic call to pd.concat would accept both of these pd.DataFrame objects in a list. By default, this will stack the objects vertically, i.e., the first pd.DataFrame is simply stacked on top of the second.

While most of the columns in our pd.DataFrame objects overlap, df_q1 does not have a close column, whereas df_q2 does. To still make the concatenation work, pandas will include the *close* column in the result of pd.concat, assigning a missing value to the rows that came from df_q1:

```
pd.concat([df_q1, df_q2])
```

	ticker	shares	low	high	close
0	AAPL	100	50	75	<NA>
1	MSFT	80	42	62	<NA>
2	AMZN	60	100	120	<NA>
0	AAPL	80	70	80	77
1	MSFT	90	50	60	55
2	IBM	100	60	70	64
3	GE	42	30	50	44

You should also take note of the row index that pandas gives in the result. In essence, pandas takes the index values of df_q1, which range from 0–2, and then takes the index values of df_q2, which range from 0–3. When creating the new row index, pandas simply retains those values, stacking them vertically in the result. If you do not care for that behavior, you can pass in ignore_index=True to pd.concat:

```
pd.concat([df_q1, df_q2], ignore_index=True)
```

	ticker	shares	low	high	close
0	AAPL	100	50	75	<NA>
1	MSFT	80	42	62	<NA>
2	AMZN	60	100	120	<NA>
3	AAPL	80	70	80	77
4	MSFT	90	50	60	55
5	IBM	100	60	70	64
6	GE	42	30	50	44

💡 **Quick tip:** Enhance your coding experience with the **AI Code Explainer** and **Quick Copy** features. Open this book in the next-gen Packt Reader. Click the **Copy** button (**1**) to quickly copy code into your coding environment, or click the **Explain** button (**2**) to get the AI assistant to explain a block of code to you.

```
function calculate(a, b) {
  return {sum: a + b};
};
```

Copy   Explain
  1   2

🔒 **The next-gen Packt Reader** is included for free with the purchase of this book. Unlock it by scanning the QR code below or visiting `https://www.packtpub.com/unlock/9781836205876`.

Another potential issue is that we can no longer see which pd.DataFrame our records originally come from. To retain that information, we can pass through a keys= argument, providing custom labels to denote the source of our data:

```
pd.concat([df_q1, df_q2], keys=["q1", "q2"])
```

		ticker	shares	low	high	close
q1	0	AAPL	100	50	75	\<NA>
	1	MSFT	80	42	62	\<NA>
	2	AMZN	60	100	120	\<NA>
q2	0	AAPL	80	70	80	77
	1	MSFT	90	50	60	55
	2	IBM	100	60	70	64
	3	GE	42	30	50	44

pd.concat also allows you to control the direction in which things are being concatenated. Instead of the default behavior to stack vertically, we can pass axis=1 to see things stacked horizontally:

```
pd.concat([df_q1, df_q2], keys=["q1", "q2"], axis=1)
```

	q1			...	q2		
	ticker	shares	low	...	low	high	close
0	AAPL	100	50	...	70	80	77
1	MSFT	80	42	...	50	60	55
2	AMZN	60	100	...	60	70	64
3	\<NA>	\<NA>	\<NA>	...	30	50	44

4 rows × 9 columns

While this gave us back a result without error, a closer inspection of the result reveals some issues. The first two rows of data cover both AAPL and MSFT, respectively, so there is no cause for concern there. However, the third row of data shows AMZN as the Q1 ticker and IBM as the Q2 ticker – what gives?

The problem is that pandas is aligning on the values of the index, and not on any other column like ticker, which is what we are probably interested in. If we wanted pd.concat to align by the ticker, we could set that as the row index of the two pd.DataFrame objects before concatenation:

```
pd.concat([
    df_q1.set_index("ticker"),
    df_q2.set_index("ticker"),
], keys=["q1", "q2"], axis=1)
```

	q1			...	q2		
	shares	low	high	...	low	high	close
ticker							
AAPL	100	50	75	...	70	80	77
MSFT	80	42	62	...	50	60	55

```
AMZN      60          100      120      ...   <NA>    <NA>    <NA>
IBM       <NA>        <NA>     <NA>     ...   60      70      64
GE        <NA>        <NA>     <NA>     ...   30      50      44
5 rows × 7 columns
```

One last thing we might want to control about the alignment behavior is how it treats labels that appear in at least one, but not all, of the objects being concatenated. By default, pd.concat performs an "outer" join, which will take all of the index values (in our case, the ticker symbols) and show them in the output, using a missing value indicator where applicable. Passing join="inner" as an argument, by contrast, will only show index labels that appear in all of the objects being concatenated:

```
pd.concat([
    df_q1.set_index("ticker"),
    df_q2.set_index("ticker"),
], keys=["q1", "q2"], axis=1, join="inner")
```

```
          q1                      ...   q2
          shares   low    high   ...   low    high    close
ticker
AAPL      100      50     75     ...   70     80      77
MSFT      80       42     62     ...   50     60      55
2 rows × 7 columns
```

There's more...

pd.concat is an expensive operation, and should never be called from within a Python loop. If you create a bunch of pd.DataFrame objects within a loop and eventually do want to concatenate them together, you are better off storing them in a sequence first, only calling pd.concat once after the sequence has been fully populated.

We can use the IPython %%time magic function to profile the difference in approaches. Let's start with the anti-pattern of using pd.concat within a loop:

```
%%time
concatenated_dfs = df_q1
for i in range(1000):
    concatenated_dfs = pd.concat([concatenated_dfs, df_q1])

print(f"Final pd.DataFrame shape is {concatenated_dfs.shape}")
```

```
Final pd.DataFrame shape is (3003, 4)
CPU times: user 267 ms, sys: 0 ns, total: 267 ms
Wall time: 287 ms
```

This code will yield the equivalent result but follows the practice of appending to a Python list during the loop, and only calling pd.concat once at the very end:

```
%%time
df = df_q1
accumulated = [df_q1]
for i in range(1000):
    accumulated.append(df_q1)


concatenated_dfs = pd.concat(accumulated)
print(f"Final pd.DataFrame shape is {concatenated_dfs.shape}")
```

```
Final pd.DataFrame shape is (3003, 4)
CPU times: user 28.4 ms, sys: 0 ns, total: 28.4 ms
Wall time: 31 ms
```

Merging DataFrames with pd.merge

Another common task in reshaping data is referred to as *merging*, or in some cases, *joining*, with the latter term being used frequently in database terminology. Where concatenation "stacks" objects on top of or next to one another, a *merge* works by finding a common key (or set of keys) between two entities and using that to blend other columns from the entities together:

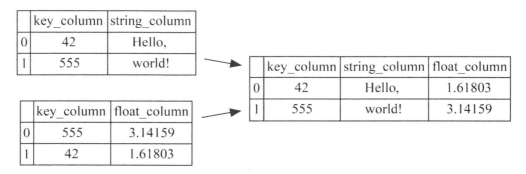

Figure 7.3: Merging two pd.DataFrame objects

The most commonly used method in pandas to perform merges is pd.merge, whose functionality will be covered throughout this recipe. Another viable, though less commonly used, pd.DataFrame.join method can be used as well, although knowing pd.merge first is helpful before discussing that (we will cover pd.DataFrame.join in the next recipe).

How to do it

Let's continue along with the stock pd.DataFrame objects we created in the *Concatenating pd.DataFrame objects* recipe:

```python
df_q1 = pd.DataFrame([
    ["AAPL", 100., 50., 75.],
    ["MSFT", 80., 42., 62.],
    ["AMZN", 60., 100., 120.],
], columns=["ticker", "shares", "low", "high"])
df_q1 = df_q1.convert_dtypes(dtype_backend="numpy_nullable")
```

```python
df_q1
```

	ticker	shares	low	high
0	AAPL	100	50	75
1	MSFT	80	42	62
2	AMZN	60	100	120

```python
df_q2 = pd.DataFrame([
    ["AAPL", 80., 70., 80., 77.],
    ["MSFT", 90., 50., 60., 55.],
    ["IBM", 100., 60., 70., 64.],
    ["GE", 42., 30., 50., 44.],
], columns=["ticker", "shares", "low", "high", "close"])
df_q2 = df_q2.convert_dtypes(dtype_backend="numpy_nullable")
```

```python
df_q2
```

	ticker	shares	low	high	close
0	AAPL	80	70	80	77
1	MSFT	90	50	60	55
2	IBM	100	60	70	64
3	GE	42	30	50	44

In one of the examples in that recipe, we saw how you could use a combination of pd.concat and pd.DataFrame.set_index to merge our two pd.DataFrame objects by the ticker column:

```python
pd.concat([
    df_q1.set_index("ticker"),
    df_q2.set_index("ticker"),
], keys=["q1", "q2"], axis=1)
```

```
        q1                          ...   q2
        shares    low     high      ...   low     high    close
ticker
AAPL    100       50      75        ...   70      80      77
MSFT    80        42      62        ...   50      60      55
AMZN    60        100     120       ...   <NA>    <NA>    <NA>
IBM     <NA>      <NA>    <NA>      ...   60      70      64
GE      <NA>      <NA>    <NA>      ...   30      50      44
5 rows × 7 columns
```

With `pd.merge`, you can express this much more succinctly by passing an argument to `on=`, which clarifies the column(s) you would like pandas to use for alignment:

```
pd.merge(df_q1, df_q2, on=["ticker"])
```

```
      ticker    shares_x    low_x    ...   low_y    high_y    close
0     AAPL      100         50       ...   70       80        77
1     MSFT      80          42       ...   50       60        55
2 rows × 8 columns
```

As you can see, the result is not exactly the same, but we can get a little closer by toggling the merge behavior. By default, `pd.merge` performs an *inner* merge; if we wanted a result more similar to our `pd.concat` example, we could pass `how="outer"`:

```
pd.merge(df_q1, df_q2, on=["ticker"], how="outer")
```

```
      ticker    shares_x    low_x    ...   low_y    high_y    close
0     AAPL      100         50       ...   70       80        77
1     AMZN      60          100      ...   <NA>     <NA>      <NA>
2     GE        <NA>        <NA>     ...   30       50        44
3     IBM       <NA>        <NA>     ...   60       70        64
4     MSFT      80          42       ...   50       60        55
5 rows × 8 columns
```

While `pd.concat` only allows you to perform *inner* or *outer* merges, `pd.merge` additionally supports *left* merges, which retain all data from the first `pd.DataFrame`, merging in data from the second `pd.DataFrame` as key fields can be matched:

```
pd.merge(df_q1, df_q2, on=["ticker"], how="left")
```

```
      ticker    shares_x    low_x    ...   low_y    high_y    close
0     AAPL      100         50       ...   70       80        77
1     MSFT      80          42       ...   50       60        55
2     AMZN      60          100      ...   <NA>     <NA>      <NA>
3 rows × 8 columns
```

`how="right"` reverses that, ensuring that every row from the second `pd.DataFrame` is represented in the output:

```
pd.merge(df_q1, df_q2, on=["ticker"], how="right")
```

```
      ticker    shares_x    low_x    ...    low_y    high_y    close
0     AAPL      100         50       ...    70       80        77
1     MSFT      80          42       ...    50       60        55
2     IBM       <NA>        <NA>     ...    60       70        64
3     GE        <NA>        <NA>     ...    30       50        44
4 rows × 8 columns
```

An additional feature when using `how="outer"` is the ability to provide an `indicator=` argument, which will tell you where each row in the resulting `pd.DataFrame` was sourced from:

```
pd.merge(df_q1, df_q2, on=["ticker"], how="outer", indicator=True)
```

```
      ticker    shares_x    low_x    ...    high_y    close    _merge
0     AAPL      100         50       ...    80        77       both
1     AMZN      60          100      ...    <NA>      <NA>     left_only
2     GE        <NA>        <NA>     ...    50        44       right_only
3     IBM       <NA>        <NA>     ...    70        64       right_only
4     MSFT      80          42       ...    60        55       both
5 rows × 9 columns
```

A value of "both" means that the key(s) used to perform the merge were found in both `pd.DataFrame` objects, which you can see is applicable to the `AAPL` and `MSFT` tickers. A value of `left_only` means the key(s) only appeared in the left `pd.DataFrame`, as is the case for `AMZN`. `right_only` highlights key(s) that only appeared in the right `pd.DataFrame`, like `GE` and `IBM`.

Another difference between our `pd.concat` output and what we get with `pd.merge` is that the former generates a `pd.MultiIndex` in the columns, essentially preventing any clashes from column labels that appear in both `pd.DataFrame` objects. `pd.merge`, by contrast, appends a suffix to columns that appear in both of the `pd.DataFrame` objects to disambiguate. The column coming from the left `pd.DataFrame` will be suffixed with `_x`, whereas a suffix of `_y` indicates that the column came from the right `pd.DataFrame`.

For more control over this suffix, you can pass a tuple argument to `suffixes=`. With our sample data, this argument can be used to easily identify Q1 versus Q2 data:

```
pd.merge(
    df_q1,
    df_q2,
    on=["ticker"],
    how="outer",
    suffixes=("_q1", "_q2"),
)
```

```
     ticker   shares_q1    low_q1    ...   low_q2   high_q2    close
0    AAPL     100          50        ...   70       80         77
1    AMZN     60           100       ...   <NA>     <NA>       <NA>
2    GE       <NA>         <NA>      ...   30       50         44
3    IBM      <NA>         <NA>      ...   60       70         64
4    MSFT     80           42        ...   50       60         55
5 rows × 8 columns
```

However, you should be aware that the suffixes are only applied if the column name appears in both `pd.DataFrame` objects. If a column only appears in one but not both objects, no suffix will be applied:

```
pd.merge(
    df_q1[["ticker"]].assign(only_in_left=42),
    df_q2[["ticker"]].assign(only_in_right=555),
    on=["ticker"],
    how="outer",
    suffixes=("_q1", "_q2"),
)
```

```
     ticker   only_in_left    only_in_right
0    AAPL     42.0            555.0
1    AMZN     42.0            NaN
2    GE       NaN             555.0
3    IBM      NaN             555.0
4    MSFT     42.0            555.0
```

If our key column(s) has different names in the two `pd.DataFrame` objects, would that be a problem? Of course not! No need to take my word for it though – let's just rename the `ticker` column in one of our `pd.DataFrame` objects to `SYMBOL`:

```
df_q2 = df_q2.rename(columns={"ticker": "SYMBOL"})
```

```
df_q2
```

```
     SYMBOL   shares    low    high    close
0    AAPL     80        70     80      77
1    MSFT     90        50     60      55
2    IBM      100       60     70      64
3    GE       42        30     50      44
```

With `pd.merge`, the only thing that changes is that you now need to pass two different arguments to `left_on=` and `right_on=`, instead of just one argument to `on=`:

```
pd.merge(
    df_q1,
    df_q2,
```

```
        left_on=["ticker"],
        right_on=["SYMBOL"],
        how="outer",
        suffixes=("_q1", "_q2"),
    )
```

	ticker	shares_q1	low_q1	...	low_q2	high_q2	close
0	AAPL	100	50	...	70	80	77
1	AMZN	60	100	...	<NA>	<NA>	<NA>
2	<NA>	<NA>	<NA>	...	30	50	44
3	<NA>	<NA>	<NA>	...	60	70	64
4	MSFT	80	42	...	50	60	55

5 rows × 9 columns

To finish off this recipe, let's consider a case where there are multiple columns that should comprise our merge key. We can start down this path by creating one pd.DataFrame that lists out the ticker, quarter, and low price:

```
lows = pd.DataFrame([
    ["AAPL", "Q1", 50.],
    ["MSFT", "Q1", 42.],
    ["AMZN", "Q1", 100.],
    ["AAPL", "Q2", 70.],
    ["MSFT", "Q2", 50.],
    ["IBM", "Q2", 60.],
    ["GE", "Q2", 30.],
], columns=["ticker", "quarter", "low"])
lows = lows.convert_dtypes(dtype_backend="numpy_nullable")

lows
```

	ticker	quarter	low
0	AAPL	Q1	50
1	MSFT	Q1	42
2	AMZN	Q1	100
3	AAPL	Q2	70
4	MSFT	Q2	50
5	IBM	Q2	60
6	GE	Q2	30

A second pd.DataFrame will also contain the ticker and quarter (albeit with different names), but will show the highs instead of the lows:

```
highs = pd.DataFrame([
    ["AAPL", "Q1", 75.],
    ["MSFT", "Q1", 62.],
    ["AMZN", "Q1", 120.],
    ["AAPL", "Q2", 80.],
    ["MSFT", "Q2", 60.],
    ["IBM", "Q2", 70.],
    ["GE", "Q2", 50.],
], columns=["SYMBOL", "QTR", "high"])
highs = highs.convert_dtypes(dtype_backend="numpy_nullable")

highs
```

	SYMBOL	QTR	high
0	AAPL	Q1	75
1	MSFT	Q1	62
2	AMZN	Q1	120
3	AAPL	Q2	80
4	MSFT	Q2	60
5	IBM	Q2	70
6	GE	Q2	50

With the layout of these pd.DataFrame objects, our key field now becomes the combination of the ticker and the quarter. By passing the appropriate labels as arguments to left_on= and right_on=, pandas is still able to perform this merge:

```
pd.merge(
    lows,
    highs,
    left_on=["ticker", "quarter"],
    right_on=["SYMBOL", "QTR"],
)
```

	ticker	quarter	low	SYMBOL	QTR	high
0	AAPL	Q1	50	AAPL	Q1	75
1	MSFT	Q1	42	MSFT	Q1	62
2	AMZN	Q1	100	AMZN	Q1	120
3	AAPL	Q2	70	AAPL	Q2	80
4	MSFT	Q2	50	MSFT	Q2	60
5	IBM	Q2	60	IBM	Q2	70
6	GE	Q2	30	GE	Q2	50

There's more...

An extra consideration when trying to merge data is the uniqueness of the key(s) in both pd.DataFrame objects. Having a poor or incorrect understanding of this can lead to very hard-to-detect errors appearing in your applications. Fortunately, pd.merge can help detect these issues upfront.

To illustrate what we mean when we talk about uniqueness, highlight the issues it can cause, and show you how to solve them with pandas, let's start with a small pd.DataFrame that shows hypothetical sales by salesperson over time:

```
sales = pd.DataFrame([
    ["Jan", "John", 10],
    ["Feb", "John", 20],
    ["Mar", "John", 30],
], columns=["month", "salesperson", "sales"])
sales = sales.convert_dtypes(dtype_backend="numpy_nullable")

sales
```

```
     month   salesperson   sales
0    Jan     John          10
1    Feb     John          20
2    Mar     John          30
```

Let's also create a separate pd.DataFrame that maps each salesperson to a particular region:

```
regions = pd.DataFrame([
    ["John", "Northeast"],
    ["Jane", "Southwest"],
], columns=["salesperson", "region"])
regions = regions.convert_dtypes(dtype_backend="numpy_nullable")

regions
```

```
     salesperson   region
0    John          Northeast
1    Jane          Southwest
```

If you have ever worked at a small company or within a small department, chances are you've seen data sources built this way. As far as employees in that space are concerned, everyone knows who John is, so they are content with the decision to lay out data in this fashion.

In the sales data, John appears multiple times, but in the regions data, John appears only once. Therefore, using salesperson as the merge key, the relationship from sales to regions is many-to-one (n-to-1). Conversely, the relationship from regions to sales is one-to-many (1-to-n).

With these types of relationships, merges do not introduce any unexpected behavior. A `pd.merge` between these two objects will simply display the multiple rows of sales data alongside the corresponding region information:

```
pd.merge(sales, regions, on=["salesperson"])
```

	month	salesperson	sales	region
0	Jan	John	10	Northeast
1	Feb	John	20	Northeast
2	Mar	John	30	Northeast

If we were to try and sum the sales of this after the merge, we would still get the appropriate amount of 60:

```
pd.merge(sales, regions, on=["salesperson"])["sales"].sum()
```

```
60
```

As the company or department grows, it becomes inevitable that another John gets hired. To accommodate this, our regions, `pd.DataFrame` gets updated to add a new `last_name` column, and add a new entry for John Newhire:

```
regions_orig = regions
regions = pd.DataFrame([
    ["John", "Smith", "Northeast"],
    ["Jane", "Doe", "Southwest"],
    ["John", "Newhire", "Southeast"],
], columns=["salesperson", "last_name", "region"])
regions = regions.convert_dtypes(dtype_backend="numpy_nullable")

regions
```

	salesperson	last_name	region
0	John	Smith	Northeast
1	Jane	Doe	Southwest
2	John	Newhire	Southeast

Suddenly, the same `pd.merge` we performed before yields a different result:

```
pd.merge(sales, regions, on=["salesperson"])
```

	month	salesperson	sales	last_name	region
0	Jan	John	10	Smith	Northeast
1	Jan	John	10	Newhire	Southeast
2	Feb	John	20	Smith	Northeast
3	Feb	John	20	Newhire	Southeast
4	Mar	John	30	Smith	Northeast
5	Mar	John	30	Newhire	Southeast

This is a definite programming mistake. If you were to try and sum the sales column from the merged pd.DataFrame, you would end up doubling the true amount of things that were actually sold. In sum, we only sold 60 units, but with the introduction of John Newhire into our regions, pd.DataFrame suddenly changed the relationship between the two pd.DataFrame objects to many-to-many (or *n*-to-*n*), which duplicates much of our data and yields the wrong number of sales:

```
pd.merge(sales, regions, on=["salesperson"])["sales"].sum()
```

```
120
```

To catch these surprises upfront with pandas, you can provide a validate= argument to pd.merge, which establishes the expected relationship of the merge key between the two objects. A validation of many_to_one with our original pd.DataFrame objects would have been fine:

```
pd.merge(sales, regions_orig, on=["salesperson"], validate="many_to_one")
```

```
    month  salesperson  sales  region
0   Jan    John         10     Northeast
1   Feb    John         20     Northeast
2   Mar    John         30     Northeast
```

Yet that same validation would have thrown an error when John Newhire made his way into our merge:

```
pd.merge(sales, regions, on=["salesperson"], validate="many_to_one")
```

```
MergeError: Merge keys are not unique in right dataset; not a many-to-one merge
```

In this simplistic example, we could have avoided this issue by modeling our data differently upfront, by either using a natural key comprising multiple columns in our sales pd.DataFrame or by opting for surrogate keys in both pd.DataFrame objects. Because these examples were so small, we could have also visually identified that there was a problem with our structure.

In the real world, detecting issues like this is not so simple. You may be trying to merge thousands or millions of rows of data, so even if a large number of rows were affected by relationship issues, they could be easily overlooked. Attempting to detect issues like this by hand is akin to finding a needle in a haystack, so I strongly advise using this data validation feature to avoid surprises.

While a failure is less than ideal, in this case, you have *failed loudly* and can easily identify where your modeling assumptions went wrong. Without these checks, your users will *silently* see incorrect data, which is, more often than not, a worse outcome.

Joining DataFrames with pd.DataFrame.join

While pd.merge is the most common approach for merging two different pd.DataFrame objects, the lesser used yet functionally similar pd.DataFrame.join method is another viable option. Stylistically, you can think of pd.DataFrame.join as a shortcut for when you want to augment an existing pd.DataFrame with a few more columns; by contrast, pd.merge defaults to treating both pd.DataFrame objects with equal importance.

How to do it

To drive home the point about pd.DataFrame.join being a shortcut to augment an existing pd.DataFrame, let's imagine a sales table where the row index corresponds to a salesperson but uses a surrogate key instead of a natural key:

```python
sales = pd.DataFrame(
    [[1000], [2000], [4000]],
    columns=["sales"],
    index=pd.Index([42, 555, 9000], name="salesperson_id")
)
sales = sales.convert_dtypes(dtype_backend="numpy_nullable")
sales
```

```
                sales
salesperson_id
42              1000
555             2000
9000            4000
```

Let's also then consider a dedicated pd.DataFrame that stores the metadata for some (but not all) of these salespeople:

```python
salesperson = pd.DataFrame([
    ["John", "Smith"],
    ["Jane", "Doe"],
], columns=["first_name", "last_name"], index=pd.Index(
    [555, 42], name="salesperson_id"
))
salesperson = salesperson.convert_dtypes(dtype_backend="numpy_nullable")
salesperson
```

```
                first_name    last_name
salesperson_id
555             John          Smith
42              Jane          Doe
```

Since the data we want to use to join these two pd.DataFrame objects together is in the row index, you would have to write out left_index=True and right_index=True while calling pd.merge. Also note that, because we have a salesperson_id of 9000 in our sales pd.DataFrame but no corresponding entry in salesperson, you would have to use how="left" to make sure records are not lost during the merge:

```python
pd.merge(sales, salesperson, left_index=True, right_index=True, how="left")
```

```
                sales    first_name    last_name
salesperson_id
```

```
42       1000      Jane      Doe
555      2000      John      Smith
9000     4000      <NA>      <NA>
```

That rather lengthy call to pd.merge describes the default behavior of pd.DataFrame.join, so you may find it easier just to use the latter:

```
sales.join(salesperson)
```

```
                sales    first_name    last_name
salesperson_id
42              1000     Jane          Doe
555             2000     John          Smith
9000            4000     <NA>          <NA>
```

While pd.DataFrame.join defaults to a left join, you can also choose a different behavior through the argument to how=:

```
sales.join(salesperson, how="inner")
```

```
                sales    first_name    last_name
salesperson_id
42              1000     Jane          Doe
555             2000     John          Smith
```

Ultimately, there is no requirement to use pd.DataFrame.join over pd.merge. The former is simply a shortcut and a stylistic indication that the calling pd.DataFrame (here, sales) should not drop any records when being joined against another pd.DataFrame, like salesperson.

Reshaping with pd.DataFrame.stack and pd.DataFrame. unstack

Before we jump into the terms *stacking* and *unstacking*, let's take a step back and compare two tables of data. Do you notice anything different about:

	a	b	c
x	1	2	3
y	4	5	6

Table 7.1: A table in wide format

compared to:

x	a	1
x	b	2
x	c	3
y	a	4
y	b	5
y	c	6

Table 7.2: A table in long format

Of course, visually, the tables have different shapes, but the data contained in each is the same. The former table would commonly be referred to as a table in *wide* format, as it stores data strewn across different columns. By contrast, in the second table, which many would say is stored in the *long* format, new rows are used to represent the various bits of data.

Which format is better? The answer to this is *it depends* – namely, on your audience and/or the systems you interact with. An executive at your company may prefer to see data stored in the wide format, as it is easier to read at a glance. A columnar database would prefer the long format, as it can better optimize for millions and billions of rows than it could for an equal number of columns.

Knowing that there is no single way to store data, you will likely need to reshape data in and out of both of these formats, which brings us to the terms *stacking* and *unstacking*.

Stacking refers to the process of taking your columns and pushing them down into the rows, essentially, helping to move from a wide format into a long format:

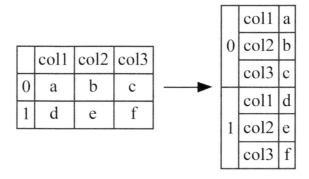

Figure 7.4: Stacking a pd.DataFrame from wide to long format

Unstacking goes in the opposite direction, moving data that is stored in a long format into a wide format:

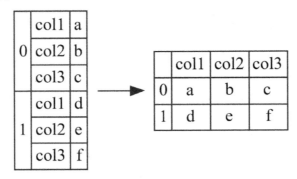

Figure 7.5: Unstacking a pd.DataFrame from long to wide format

In this recipe, we will walk you through the proper usage of the pd.DataFrame.stack and pd.DataFrame.unstack methods, which can be used for these reshaping purposes.

How to do it

Let's start with the following pd.DataFrame, which summarizes the amount of fruits being grown in different states:

```
df = pd.DataFrame([
    [12, 10, 40],
    [9, 7, 12],
    [0, 14, 190]
], columns=pd.Index(["Apple", "Orange", "Banana"], name="fruit"), index=pd.
Index(
    ["Texas", "Arizona", "Florida"], name="state"))

df = df.convert_dtypes(dtype_backend="numpy_nullable")

df
```

```
fruit    Apple   Orange  Banana
state
Texas    12      10      40
Arizona  9       7       12
Florida  0       14      190
```

In data modeling terminology, we would consider this to be a "wide" table. Each row represents one state with the different numbers of each crop situated in its own column.

If we wanted to convert our table to "long" form, we would essentially want to see each `state` and `fruit` combination as a separate row. `pd.DataFrame.stack` will help us do this, by taking our fruits out of the column index and forming a new `pd.MultiIndex` in our rows, which contains both state and fruit:

```
df.stack()
```

```
state    fruit
Texas    Apple      12
         Orange     10
         Banana     40
Arizona  Apple       9
         Orange      7
         Banana     12
Florida  Apple       0
         Orange     14
         Banana    190
dtype: Int64
```

After a call to `pd.DataFrame.stack`, many users will chain in a call to `pd.Series.reset_index` with a `name=` argument. This converts the `pd.Series` with a `pd.MultiIndex` created from the `pd.DataFrame.stack` back into a `pd.DataFrame` with meaningful column names:

```
df.stack().reset_index(name="number_grown")
```

```
   state    fruit   number_grown
0  Texas    Apple    12
1  Texas    Orange   10
2  Texas    Banana   40
3  Arizona  Apple    9
4  Arizona  Orange   7
5  Arizona  Banana   12
6  Florida  Apple    0
7  Florida  Orange   14
8  Florida  Banana   190
```

This long form of storing data is preferred for storage by many databases and is also the expected shape of the `pd.DataFrame` to be passed to libraries like Seaborn, which we showcased in the *Seaborn introduction* recipe back in *Chapter 6, Visualization*.

However, sometimes, you may want to go in the opposite direction, converting your long `pd.DataFrame` into a wider format. This can be particularly useful when wanting to summarize data in a compact area; utilizing both dimensions for display is more effective than asking your viewer to scroll through many lines of data.

To see this in action, let's create a new `pd.Series` from one of the `pd.DataFrame.stack` calls we just made:

```
stacked = df.stack()
stacked
```

```
state    fruit
Texas    Apple      12
         Orange     10
         Banana     40
Arizona  Apple       9
         Orange      7
         Banana     12
Florida  Apple       0
         Orange     14
         Banana    190
dtype: Int64
```

To go in the opposite direction and move one of our index levels from the rows to the columns, you simply need to make a call to `pd.Series.unstack`:

```
stacked.unstack()
```

```
fruit    Apple   Orange   Banana
state
Texas    12      10       40
Arizona  9       7        12
Florida  0       14       190
```

By default, a call to `pd.Series.unstack` moves the innermost level of the row index, which, in our case, was the `fruit`. However, we could have passed `level=0` to have it take the very first level instead of the innermost, in the case that we wanted to see the states summarized across the columns:

```
stacked.unstack(level=0)
```

```
state    Texas   Arizona   Florida
fruit
Apple    12      9         0
Orange   10      7         14
Banana   40      12        190
```

Because our `pd.MultiIndex` levels have names, we could also have referred to the level we wanted to be moved by name instead of by position:

```
stacked.unstack(level="state")
```

```
state   Texas   Arizona   Florida
fruit
Apple   12      9         0
Orange  10      7         14
Banana  40      12        190
```

Reshaping with pd.DataFrame.melt

In the *Reshaping with pd.DataFrame.stack and pd.DataFrame.unstack* recipe, we discovered that you could convert a wide pd.DataFrame into long form by setting the appropriate row and column index-(es) before calling pd.DataFrame.stack. pd.TheDataFrame.melt function also lets you convert your pd.DataFrame from wide to long, but can do so without having to set the row and column index values in an intermediate step, while also offering more control over what other columns may or may not be included as part of the wide to long conversion.

How to do it

Let's once again create a summary of the different fruits being grown in different states. However, unlike the *Reshaping with pd.DataFrame.stack and pd.DataFrame.unstack* recipe, we will not be setting the row index to the state values, and instead, just treating it as another column in our pd.DataFrame:

```
df = pd.DataFrame([
    ["Texas", 12, 10, 40],
    ["Arizona", 9, 7, 12],
    ["Florida", 0, 14, 190]
], columns=["state", "apple", "orange", "banana"])
df = df.convert_dtypes(dtype_backend="numpy_nullable")

df
```

	state	apple	orange	banana
0	Texas	12	10	40
1	Arizona	9	7	12
2	Florida	0	14	190

To convert to long format with pd.DataFrame.stack, we would have to chain together a few calls to get back a pd.DataFrame without a pd.MultiIndex:

```
df.set_index("state").stack().reset_index()
```

	state	level_1	0
0	Texas	apple	12
1	Texas	orange	10
2	Texas	banana	40
3	Arizona	apple	9
4	Arizona	orange	7

5	Arizona	banana	12
6	Florida	apple	0
7	Florida	orange	14
8	Florida	banana	190

The column name `level_1` is created by default during our `pd.DataFrame.stack` operation because the column index we start with is unnamed. We also see that we get an auto-generated column name of 0 for the newly introduced values in our long format, so we would still need to chain in a rename to get us a more readable `pd.DataFrame`:

```
df.set_index("state").stack().reset_index().rename(columns={
    "level_1": "fruit",
    0: "number_grown",
})
```

	state	fruit	number_grown
0	Texas	apple	12
1	Texas	orange	10
2	Texas	banana	40
3	Arizona	apple	9
4	Arizona	orange	7
5	Arizona	banana	12
6	Florida	apple	0
7	Florida	orange	14
8	Florida	banana	190

`pd.DataFrame.melt` gets us a lot closer to our desired `pd.DataFrame`, simply by providing an `id_vars=` argument that corresponds to the row index you would have used with `pd.DataFrame.stack`:

```
df.melt(id_vars=["state"])
```

	state	variable	value
0	Texas	apple	12
1	Arizona	apple	9
2	Florida	apple	0
3	Texas	orange	10
4	Arizona	orange	7
5	Florida	orange	14
6	Texas	banana	40
7	Arizona	banana	12
8	Florida	banana	190

With `pd.DataFrame.melt`, the newly created column from our variables (here, the different fruits) is given the name `variable`, and the value column is given the default name of `value`. We can override these defaults through the use of the `var_name=` and `value_name=` arguments:

```
df.melt(
    id_vars=["state"],
    var_name="fruit",
    value_name="number_grown",
)
```

```
   state    fruit    number_grown
0  Texas    apple    12
1  Arizona  apple    9
2  Florida  apple    0
3  Texas    orange   10
4  Arizona  orange   7
5  Florida  orange   14
6  Texas    banana   40
7  Arizona  banana   12
8  Florida  banana   190
```

As an added bonus, `pd.DataFrame.melt` gives you an easy way to control which columns are included as part of the wide-to-long conversion. For instance, if we don't care to include the banana values in our newly formed long table, we could just pass the other columns of `apple` and `orange` as arguments to `value_vars=`:

```
df.melt(
    id_vars=["state"],
    var_name="fruit",
    value_name="number_grown",
    value_vars=["apple", "orange"],
)
```

```
   state    fruit    number_grown
0  Texas    apple    12
1  Arizona  apple    9
2  Florida  apple    0
3  Texas    orange   10
4  Arizona  orange   7
5  Florida  orange   14
```

Reshaping with pd.wide_to_long

So far, we have encountered two very viable ways of converting data from wide to long format, whether it be through the use of the pd.DataFrame.stack method, introduced in our *Reshaping with pd.DataFrame.stack and pd.DataFrame.unstack* recipe, or through the use of pd.DataFrame.melt, as we saw in the *Reshaping with pd.DataFrame.melt* recipe.

If those aren't enough, pandas offers the pd.wide_to_long function, which can help with that conversion given that your columns follow a particular naming pattern, as we will see in this recipe.

How to do it

Let's assume we have the following pd.DataFrame, where we have one id variable of widget and four columns representing sales from a business quarter. Each column of sales begins with "quarter_":

```python
df = pd.DataFrame([
    ["Widget 1", 1, 2, 4, 8],
    ["Widget 2", 16, 32, 64, 128],
], columns=["widget", "quarter_1", "quarter_2", "quarter_3", "quarter_4"])
df = df.convert_dtypes(dtype_backend="numpy_nullable")

df
```

	widget	quarter_1	quarter_2	quarter_3	quarter_4
0	Widget 1	1	2	4	8
1	Widget 2	16	32	64	128

Going back to our example of pd.DataFrame.stack, we could convert this from wide to long using the following methods:

```python
df.set_index("widget").stack().reset_index().rename(columns={
    "level_1": "quarter",
    0: "quantity",
})
```

	widget	quarter	quantity
0	Widget 1	quarter_1	1
1	Widget 1	quarter_2	2
2	Widget 1	quarter_3	4
3	Widget 1	quarter_4	8
4	Widget 2	quarter_1	16
5	Widget 2	quarter_2	32
6	Widget 2	quarter_3	64
7	Widget 2	quarter_4	128

For a more succinct solution, we could use pd.DataFrame.melt:

```
df.melt(
    id_vars=["widget"],
    var_name="quarter",
    value_name="quantity",
)
```

```
      widget      quarter      quantity
0     Widget 1    quarter_1    1
1     Widget 2    quarter_1    16
2     Widget 1    quarter_2    2
3     Widget 2    quarter_2    32
4     Widget 1    quarter_3    4
5     Widget 2    quarter_3    64
6     Widget 1    quarter_4    8
7     Widget 2    quarter_4    128
```

But there is a feature that pd.wide_to_long offers that neither of these approaches handles directly – namely, to create a new variable out of the column labels that are being converted into variables. So far, we see the new quarter values as quarter_1, quarter_2, quarter_3, and quarter_4, but pd.wide_to_long can extract that string out of the newly created variables, more simply leaving you with the digits 1, 2, 3, and 4:

```
pd.wide_to_long(
    df,
    i=["widget"],
    stubnames="quarter_",
    j="quarter"
).reset_index().rename(columns={"quarter_": "quantity"})
```

```
      widget      quarter    quantity
0     Widget 1    1          1
1     Widget 2    1          16
2     Widget 1    2          2
3     Widget 2    2          32
4     Widget 1    3          4
5     Widget 2    3          64
6     Widget 1    4          8
7     Widget 2    4          128
```

Reshaping with pd.DataFrame.pivot and pd.pivot_table

So far in this chapter, we have seen that pd.DataFrame.stack, pd.DataFrame.melt, and pd.wide_to_long can all be used to help you convert your pd.DataFrame from a wide to a long format. On the flip side, we have only seen pd.Series.unstack helps us go from long to wide, but that method has the downside of requiring us to assign a proper row index before we can use it. With pd.DataFrame.pivot, you can skip any intermediate steps and go directly from a long to a wide format.

Beyond pd.DataFrame.pivot, pandas offers a pd.pivot_table function, which can not only reshape from long to wide but allows you to perform aggregations as part of the reshape.

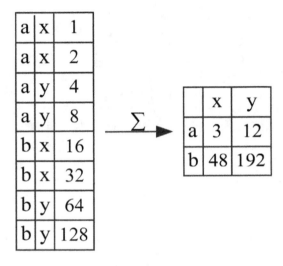

Figure 7.6: Using pd.pivot_table to reshape with sum aggregation

Effective use of pd.pivot_table allows you to perform very complex calculations with a compact and concise syntax.

How to do it

In many of the preceding recipes, we have started with data in wide form and reshaped it to long form. For this recipe, let's start with data that appears in long form from the outset. We are also going to add a new column for number_eaten to showcase the aggregation capabilities when pivoting within pandas:

```
df = pd.DataFrame([
    ["Texas", "apple", 12, 8],
    ["Arizona", "apple", 9, 10],
    ["Florida", "apple", 0, 6],
    ["Texas", "orange", 10, 4],
    ["Arizona", "orange", 7, 2],
    ["Florida", "orange", 14, 3],
    ["Texas", "banana", 40, 28],
    ["Arizona", "banana", 12, 17],
```

```
        ["Florida", "banana", 190, 42],
    ], columns=["state", "fruit", "number_grown", "number_eaten"])
    df = df.convert_dtypes(dtype_backend="numpy_nullable")

    df
```

	state	fruit	number_grown	number_eaten
0	Texas	apple	12	8
1	Arizona	apple	9	10
2	Florida	apple	0	6
3	Texas	orange	10	4
4	Arizona	orange	7	2
5	Florida	orange	14	3
6	Texas	banana	40	28
7	Arizona	banana	12	17
8	Florida	banana	190	42

As we learned back in the *Reshaping with pd.DataFrame.stack and pd.DataFrame.unstack* recipe, if we wanted to convert this from long format into wide, we could do so with the clever use of pd.DataFrame. set_index paired with pd.DataFrame.unstack:

```
    df.set_index(["state", "fruit"]).unstack()
```

	number_grown			number_eaten		
fruit	apple	banana	orange	apple	banana	orange
state						
Arizona	9	12	7	10	17	2
Florida	0	190	14	6	42	3
Texas	12	40	10	8	28	4

pd.DataFrame.pivot lets us tackle this in one method call. A basic usage of this method requires index= and columns= arguments, to dictate which column(s) should appear in the row and column indexes, respectively:

```
    df.pivot(index=["state"], columns=["fruit"])
```

	number_grown			number_eaten		
fruit	apple	banana	orange	apple	banana	orange
state						
Arizona	9	12	7	10	17	2
Florida	0	190	14	6	42	3
Texas	12	40	10	8	28	4

pd.DataFrame.pivot will take any column that is not specified as an argument to index= or columns=, and try to convert that column into the values of the resulting pd.DataFrame. However, if you did not care for all of the remaining columns to be a part of the pivoted pd.DataFrame, you could specify what you want to keep with the values= argument. For example, if we only cared to pivot the number_grown column and ignore the number_eaten column, we could write this as:

```
df.pivot(
    index=["state"],
    columns=["fruit"],
    values=["number_grown"],
)
```

```
        number_grown
fruit   apple   banana   orange
state
Arizona 9       12       7
Florida 0       190      14
Texas   12      40       10
```

In the case where you only wanted to keep one value, the generated pd.MultiIndex in the columns may seem superfluous. Fortunately, this can be dropped with a simple call to pd.DataFrame.droplevel, where you indicate the axis= where you would like to drop a level (specify 1 for the columns) and the index level you would like to drop (here, 0 represents the first level):

```
wide_df = df.pivot(
    index=["state"],
    columns=["fruit"],
    values=["number_grown"],
).droplevel(level=0, axis=1)

wide_df
```

```
fruit   apple   banana   orange
state
Arizona 9       12       7
Florida 0       190      14
Texas   12      40       10
```

While pd.DataFrame.pivot is useful for reshaping, it can only help in the case that none of the values used to form your rows and columns are duplicated. To see this limitation, let's work with a slightly modified pd.DataFrame that shows how different fruits have been consumed or grown in different states and years:

```
df = pd.DataFrame([
    ["Texas", "apple", 2023, 10, 6],
```

```
        ["Texas", "apple", 2024, 2, 8],
        ["Arizona", "apple", 2023, 3, 7],
        ["Arizona", "apple", 2024, 6, 3],
        ["Texas", "orange", 2023, 5, 2],
        ["Texas", "orange", 2024, 5, 2],
        ["Arizona", "orange", 2023, 7, 2],
    ], columns=["state", "fruit", "year", "number_grown", "number_eaten"])
    df = df.convert_dtypes(dtype_backend="numpy_nullable")

    df
```

	state	fruit	year	number_grown	number_eaten
0	Texas	apple	2023	10	6
1	Texas	apple	2024	2	8
2	Arizona	apple	2023	3	7
3	Arizona	apple	2024	6	3
4	Texas	orange	2023	5	2
5	Texas	orange	2024	5	2
6	Arizona	orange	2023	7	2

We would be able to still use pd.DataFrame.pivot on this pd.DataFrame if we placed state, fruit, and year all in either the rows or the columns:

```
df.pivot(
    index=["state", "year"],
    columns=["fruit"],
    values=["number_grown", "number_eaten"]
)
```

		number_grown		number_eaten	
	fruit	apple	orange	apple	orange
state	year				
Arizona	2023	3	7	7	2
	2024	6	NaN	3	NaN
Texas	2023	10	5	6	2
	2024	2	5	8	2

But what if we didn't want to see the year as part of our output? Just removing it from our pd.DataFrame. pivot arguments will raise an exception:

```
df.pivot(
    index=["state"],
    columns=["fruit"],
    values=["number_grown", "number_eaten"]
)
```

```
ValueError: Index contains duplicate entries, cannot reshape
```

For pd.pivot_table, the lack of a year column is no problem at all:

```
pd.pivot_table(
    df,
    index=["state"],
    columns=["fruit"],
    values=["number_grown", "number_eaten"]
)
```

	number_eaten		number_grown	
fruit	apple	orange	apple	orange
state				
Arizona	5.0	2.0	4.5	7.0
Texas	7.0	2.0	6.0	5.0

This works because pd.pivot_table aggregates the values while reshaping them into a wide form. Taking Arizona apples as an example, our input data showed that a whopping three were grown in the year 2023 before doubling to a magnificent six in 2024. In our call to pd.pivot_table, this is shown as 4.5. By default, pd.pivot_table will take the average of values you supply to it during a reshape.

You can, of course, control the aggregation function being used. In this particular case, we may be more interested in knowing how many fruits were grown in each state in total, rather than taking an average by year. By passing a different aggregation function as a parameter to aggfunc=, you can easily get a summation instead:

```
pd.pivot_table(
    df,
    index=["state"],
    columns=["fruit"],
    values=["number_grown", "number_eaten"],
    aggfunc="sum"
)
```

	number_eaten		number_grown	
fruit	apple	orange	apple	orange
state				
Arizona	10	2	9	7
Texas	14	4	12	10

For more advanced use cases, you can even provide a dictionary of values to aggfunc=, where each key/value pair of the dictionary dictates the column and the type of aggregation(s) to be applied, respectively:

```
pd.pivot_table(
```

```
    df,
    index=["state"],
    columns=["fruit"],
    values=["number_grown", "number_eaten"],
    aggfunc={
        "number_eaten": ["min", "max"],
        "number_grown": ["sum", "mean"],
    },
)
```

```
        number_eaten              ...   number_grown
        max                 min   ...   mean    sum
fruit   apple    orange     apple ...   orange  apple   orange
state
Arizona 7        2          3     ...   7.0     9       7
Texas   8        2          6     ...   5.0     12      10
2 rows × 8 columns
```

Reshaping with pd.DataFrame.explode

The world would be so simple if every piece of data fitted perfectly as a scalar into a two-dimensional pd.DataFrame. Alas, life is not so simple. Especially when working with semi-structured sources of data like JSON, it is not uncommon to have individual items in your pd.DataFrame contain non-scalar sequences like lists and tuples.

You may find it acceptable to leave data in that state, but other times, there is value to normalizing the data and potentially extracting out sequences contained within a column into individual elements.

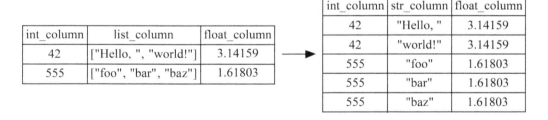

Figure 7.7: Using pd.DataFrame.explode to extract list elements to individual rows

To that end, pd.DataFrame.explode is the right tool for the job. It may not be a function you use every day, but when you eventually need to use it, you will be happy to have known about it. Attempting to replicate the same functionality outside of pandas can be error-prone and non-performant!

How to do it

Since we mentioned JSON as a good source for semi-structured data in the introduction to this recipe, let's start by imagining that we have to interact with a REST API for an HR system. The HR system should tell us who each person is in the company, as well as who, if anyone, reports to them.

The hierarchy between employees is rather easy to represent in a semi-structured format like JSON, so the REST API might return something like:

```
[
    {
        "employee_id": 1,
        "first_name": "John",
        "last_name": "Smith",
        "direct_reports": [2, 3]
    },
    {
        "employee_id": 2,
        "first_name": "Jane",
        "last_name": "Doe",
        "direct_reports": []
    },
    {
        "employee_id": 3,
        "first_name": "Joe",
        "last_name": "Schmoe",
        "direct_reports": []
    }
]
```

The pandas library will also let us load this data into a `pd.DataFrame`, albeit with the `direct_reports` column containing lists:

```
df = pd.DataFrame(
    [
        {
            "employee_id": 1,
            "first_name": "John",
            "last_name": "Smith",
            "direct_reports": [2, 3]
        },
        {
            "employee_id": 2,
```

```
            "first_name": "Jane",
            "last_name": "Doe",
            "direct_reports": []
        },
        {
            "employee_id": 3,
            "first_name": "Joe",
            "last_name": "Schmoe",
            "direct_reports": []
        }
    ]
)
df = df.convert_dtypes(dtype_backend="numpy_nullable")

df
```

	employee_id	first_name	last_name	direct_reports
0	1	John	Smith	[2, 3]
1	2	Jane	Doe	[]
2	3	Joe	Schmoe	[]

With pd.DataFrame.explode, you can unpack those direct_reports into separate rows of the pd.DataFrame:

```
df.explode("direct_reports").convert_dtypes(dtype_backend="numpy_nullable")
```

	employee_id	first_name	last_name	direct_reports
0	1	John	Smith	2
0	1	John	Smith	3
1	2	Jane	Doe	<NA>
2	3	Joe	Schmoe	<NA>

Building off of the knowledge we picked up about merging/joining data from our *Merging DataFrames with pd.merge* recipe, we can very easily take our exploded information and merge in the names of direct reports, yielding an easy summary of who works at the company and who, if anyone, reports to them:

```
exploded = df.explode("direct_reports").convert_dtypes(
    dtype_backend="numpy_nullable"
)
pd.merge(
    exploded,
    df.drop(columns=["direct_reports"]),
    how="left",
    left_on=["direct_reports"],
```

```
    right_on=["employee_id"],
    suffixes=("", "_direct_report"),
)
```

```
   employee_id  first_name  last_name  ...  employee_id_direct_report  first_
name_direct_report  last_name_direct_report
0  1            John        Smith      ...  2                          Jane                Doe
1  1            John        Smith      ...  3                          Joe                 Schmoe
2  2            Jane        Doe        ...  <NA>         <NA>           <NA>
3  3            Joe         Schmoe     ...  <NA>         <NA>           <NA>
4 rows × 7 columns
```

There's more...

While we did not introduce it in our review of types in *Chapter 3, Data Types*, PyArrow does offer a struct data type which, when used in a pd.Series, exposes a pd.Series.struct.explode method:

```python
dtype = pd.ArrowDtype(pa.struct([
    ("int_col", pa.int64()),
    ("str_col", pa.string()),
    ("float_col", pa.float64()),
]))
ser = pd.Series([
    {"int_col": 42, "str_col": "Hello, ", "float_col": 3.14159},
    {"int_col": 555, "str_col": "world!", "float_col": 3.14159},
], dtype=dtype)
ser
```

```
0    {'int_col': 42, 'str_col': 'Hello, ', 'float_c...
1    {'int_col': 555, 'str_col': 'world!', 'float_c...
dtype: struct<int_col: int64, str_col: string, float_col: double>[pyarrow]
```

Unlike pd.DataFrame.explode, which generates new rows of data, pd.Series.struct.explode generates new columns of data from its struct members:

```python
ser.struct.explode()
```

```
   int_col  str_col  float_col
0  42       Hello,   3.14159
1  555      world!   3.14159
```

This could be particularly useful if you are dealing with a semi-structured data source like JSON. If you are able to fit nested data from such a source into the typed struct that PyArrow has to offer, pd.Series.struct.explode can save you a significant amount of trouble when trying to unnest that data.

Transposing with pd.DataFrame.T

For the final recipe in this chapter, let's explore one of the easier reshaping features of pandas. *Transposition* refers to the process of inverting your pd.DataFrame so that the rows become the columns and the columns become the rows:

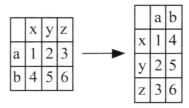

Figure 7.8: Transposing a pd.DataFrame

In this recipe, we will see how to transpose with the pd.DataFrame.T method while discussing how this might be useful.

How to do it

Transposition in pandas is straightforward. Take any pd.DataFrame:

```
df = pd.DataFrame([
    [1, 2, 3],
    [4, 5, 6],
], columns=list("xyz"), index=list("ab"))

df
```

```
     x    y    z
a    1    2    3
b    4    5    6
```

You can simply access the pd.DataFrame.T attribute and watch as your rows become your columns and your columns become your rows:

```
df.T
```

```
     a    b
x    1    4
y    2    5
z    3    6
```

There are an endless number of reasons why you may want to transpose, ranging from simply thinking *it looks better* in a given format to cases where you find it easier to select by a row index label instead of by a column index label.

However, one of the main use cases to transpose will be to get your pd.DataFrame in an *optimal* format before applying functions. As we learned back in *Chapter 5, Algorithms and How to Apply Them*, pandas has the ability to apply aggregations to each column:

```
df.sum()
```

```
x    5
y    7
z    9
dtype: int64
```

as well as to each row with the axis=1 argument:

```
df.sum(axis=1)
```

```
a     6
b    15
dtype: int64
```

Unfortunately, using the axis=1 argument can drastically reduce the performance of your applications. If you find yourself scattering a lot of axis=1 calls throughout your code, chances are you would be much better off transposing your data first and then applying functions with the default axis=0.

To see the difference, let's look at a pd.DataFrame that is rather wide:

```
np.random.seed(42)
df = pd.DataFrame(
    np.random.randint(10, size=(2, 10_000)),
    index=list("ab"),
)

df
```

```
   0   1   2  ...  9997  9998  9999
a  6   3   7  ...     2     9     4
b  2   4   2  ...     1     5     5
2 rows × 10,000 columns
```

Ultimately, we will get the same result, whether we sum the rows:

```
df.sum(axis=1)
```

```
a    44972
b    45097
dtype: int64
```

or transpose first, and then use the default summation of the columns:

```
df.T.sum()
```

```
a    44972
b    45097
dtype: int64
```

However, if you were to repeatedly make calls using axis=1 as an argument, you would find that transposing first can save significant time.

To measure this, let's use IPython and check how long it takes to perform our sum 100 times:

```
import timeit


def baseline_sum():
    for _ in range(100):
        df.sum(axis=1)


timeit.timeit(baseline_sum, number=100)
```

```
4.366703154002607
```

Comparatively, transposing first and then performing the sum will be much faster:

```
def transposed_sum():
    transposed = df.T
    for _ in range(100):
        transposed.sum()


timeit.timeit(transposed_sum, number=100)
```

```
0.7069798299999093
```

Overall, using pd.DataFrame.T to avoid subsequent calls with axis=1 is a highly encouraged practice.

Join our community on Discord

Join our community's Discord space for discussions with the authors and other readers:

```
https://packt.link/pandas
```

8

Group By

One of the most fundamental tasks during data analysis involves splitting data into independent groups before performing a calculation on each group. This methodology has been around for quite some time, but has more recently been referred to as *split-apply-combine*.

Within the *apply* step of the *split-apply-combine* paradigm, it is additionally helpful to know whether we are trying to perform a *reduction* (also referred to as an aggregation) or a *transformation*. The former reduces the values in a group down to *one value* whereas the latter attempts to maintain the shape of the group.

To illustrate, here is what *split-apply-combine* looks like for a reduction:

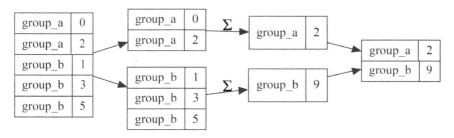

Figure 8.1: Split-apply-combine paradigm for a reduction

Here is the same paradigm for a *transformation*:

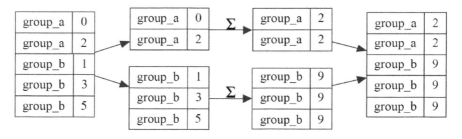

Figure 8.2: Split-apply-combine paradigm for a transformation

In pandas, the `pd.DataFrame.groupby` method is responsible for splitting, applying a function of your choice, and combining the results back together for you as an end user.

We are going to cover the following recipes in this chapter:

- Group by basics
- Grouping and calculating multiple columns
- Group by apply
- Window operations
- Selecting the highest rated movies by year
- Comparing the best hitter in baseball across years

Group by basics

True mastery of the pandas group by mechanisms is a powerful skill for any data analyst. With pandas, you can easily summarize data, find patterns within different groups, and compare groups to one another. The number of algorithms you can apply alongside a group by are endless in theory, giving you as an analyst tons of flexibility to explore your data.

In this first recipe, we are going to start with a very simple summation against different groups in an intentionally small dataset. While this example is overly simplistic, a solid theoretical understanding of how group by works is important as you look toward real-world applications.

How to do it

To get familiarized with how group by works in code, let's create some sample data that matches our starting point in *Figures 8.1* and *8.2*:

```
df = pd.DataFrame([
    ["group_a", 0],
    ["group_a", 2],
    ["group_b", 1],
    ["group_b", 3],
    ["group_b", 5],
], columns=["group", "value"])
df = df.convert_dtypes(dtype_backend="numpy_nullable")

df
```

```
    group      value
0   group_a    0
1   group_a    2
2   group_b    1
3   group_b    3
4   group_b    5
```

Our pd.DataFrame has two distinct groups: group_a and group_b. As you can see, the group_a rows are associated with value data of 0 and 2, whereas the group_b rows are associated with value data of 1, 3, and 5. Summing the values within each group should therefore yield a result of 2 and 9, respectively.

To express this with pandas, you are going to use the pd.DataFrame.groupby method, which accepts as an argument the group name(s). In our case, this is the group column. This technically returns a pd.core.groupby.DataFrameGroupBy object that exposes a pd.core.groupby.DataFrameGroupBy.sum method for summation:

```
df.groupby("group").sum()
```

group	value
group_a	2
group_b	9

Don't worry if you find the method name pd.core.groupby.DataFrameGroupBy.sum to be verbose; it is, but you will never have to write it out by hand. We are going to refer to it here by its technical name for the sake of completeness, but as an end user, you will always follow the form you can see here:

```
df.groupby(<GROUP_OR_GROUPS>)
```

This is what you will follow to get your pd.core.groupby.DataFrameGroupBy object.

By default, pd.core.groupby.DataFrameGroupBy.sum is considered an *aggregation*, so each group is *reduced* down to a single row during the *apply* phase of *split-apply-combine*, much like we see in *Figure 8.1*.

Instead of calling pd.core.groupby.DataFrameGroupBy.sum directly, we could have alternatively used the pd.core.groupby.DataFrameGroupBy.agg method, providing it with the argument of "sum":

```
df.groupby("group").agg("sum")
```

group	value
group_a	2
group_b	9

The explicitness of pd.core.groupby.DataFrameGroupBy.agg is useful when compared side by side with the pd.core.groupby.DataFrameGroupBy.transform method, which will perform a *transformation* (see *Figure 8.2* again) instead of a *reduction*:

```
df.groupby("group").transform("sum")
```

	value
0	2
1	2
2	9
3	9
4	9

pd.core.groupby.DataFrameGroupBy.transform guarantees to return a like-indexed object to the caller, which makes it ideal for performing calculations like % of group:

```
df[["value"]].div(df.groupby("group").transform("sum"))
```

```
      value
0     0.000000
1     1.000000
2     0.111111
3     0.333333
4     0.555556
```

When applying a reduction algorithm, pd.DataFrame.groupby will take the unique values of the group(s) and use them to form a new row pd.Index (or pd.MultiIndex, in the case of multiple groups). If you would prefer not to have the grouped labels create a new index, keeping them as columns instead, you can pass as_index=False:

```
df.groupby("group", as_index=False).sum()
```

```
      group     value
0     group_a     2
1     group_b     9
```

You should also note that the name of any non-grouping columns will not be altered when performing a group by operation. For example, even though we start with a pd.DataFrame containing a column named value...

```
df
```

```
      group     value
0     group_a     0
1     group_a     2
2     group_b     1
3     group_b     3
4     group_b     5
```

...the fact that we then group by the group column and sum the value column does not change its name in the result; it is still just value:

```
df.groupby("group").sum()
```

```
group      value
group_a    2
group_b    9
```

This can be confusing or ambiguous if you apply other algorithms to your groups, like `min`:

```
df.groupby("group").min()
```

```
group      value
group_a    0
group_b    1
```

Our column is still just called `value`, even though in one instance, we are taking the *sum of value* and in the other instance, we are taking the *min of value*.

Fortunately, there is a way to control this by using the `pd.NamedAgg` class. When calling `pd.core.groupby.DataFrameGroupBy.agg`, you can provide keyword arguments where each argument key dictates the desired column name and the argument value is a `pd.NamedAgg`, which dictates the aggregation as well as the original column it is applied to.

For instance, if we wanted to apply a `sum` aggregation to our `value` column, and have the result shown as `sum_of_value`, we could write the following:

```
df.groupby("group").agg(sum_of_value=pd.NamedAgg(column="value",
aggfunc="sum"))
```

```
group      sum_of_value
group_a    2
group_b    9
```

There's more...

Although this recipe focused mainly on summation, pandas offers quite a few other built-in *reduction* algorithms that can be applied to a `pd.core.groupby.DataFrameGroupBy` object, such as the following:

any	all	sum	prod
idxmin	idxmax	min	max
mean	median	var	std
sem	skew	first	last

Table 8.1: Commonly used GroupBy reduction algorithms

Likewise, there are some built-in *transformation* functions that you can use:

cumprod	cumsum	cummin
cummax	rank	

Table 8.2: Commonly used GroupBy transformation algorithms

Functionally, there is no difference between calling these functions directly as methods of `pd.core.groupby.DataFrameGroupBy` versus providing them as an argument to `pd.core.groupby.DataFrameGroupBy.agg` or `pd.core.groupby.DataFrameGroupBy.transform`. You will get the same performance and result by doing the following:

```
df.groupby("group").max()
```

```
group       value
group_a     2
group_b     5
```

The preceding code snippet will yield the same results as this one:

```
df.groupby("group").agg("max")
```

```
group       value
group_a     2
group_b     5
```

You could argue that the latter approach signals a clearer intent, especially considering that max can be used as a transformation just as well as an aggregation:

```
df.groupby("group").transform("max")
```

```
    value
0       2
1       2
2       5
3       5
4       5
```

In practice, both styles are commonplace, so you should be familiar with the different approaches.

Grouping and calculating multiple columns

Now that we have the basics down, let's take a look at a `pd.DataFrame` that contains more columns of data. Generally, your `pd.DataFrame` objects will contain many columns with potentially different data types, so knowing how to select and work with them all through the context of `pd.core.groupby.DataFrameGroupBy` is important.

How to do it

Let's create a `pd.DataFrame` that shows the `sales` and `returns` of a hypothetical `widget` across different `region` and `month` values:

```
df = pd.DataFrame([
    ["North", "Widget A", "Jan", 10, 2],
    ["North", "Widget B", "Jan", 4, 0],
```

```
        ["South", "Widget A", "Jan", 8, 3],
        ["South", "Widget B", "Jan", 12, 8],
        ["North", "Widget A", "Feb", 3, 0],
        ["North", "Widget B", "Feb", 7, 0],
        ["South", "Widget A", "Feb", 11, 2],
        ["South", "Widget B", "Feb", 13, 4],
    ], columns=["region", "widget", "month", "sales", "returns"])
    df = df.convert_dtypes(dtype_backend="numpy_nullable")


    df
```

```
      region    widget    month   sales    returns
0     North     Widget A  Jan     10       2
1     North     Widget B  Jan     4        0
2     South     Widget A  Jan     8        3
3     South     Widget B  Jan     12       8
4     North     Widget A  Feb     3        0
5     North     Widget B  Feb     7        0
6     South     Widget A  Feb     11       2
7     South     Widget B  Feb     13       4
```

To calculate the total `sales` and `returns` for each `widget`, your first attempt at doing so may look like this:

```
df.groupby("widget").sum()
```

```
widget      region                month          sales   returns
Widget A    NorthSouthNorthSouth  JanJanFebFeb   32      7
Widget B    NorthSouthNorthSouth  JanJanFebFeb   36      12
```

While `sales` and `returns` look good, the `region` and `month` columns also ended up being summed, using the same summation logic that Python would when working with strings:

```
"North" + "South" + "North" + "South"
```

```
NorthSouthNorthSouth
```

Unfortunately, this default behavior is usually undesirable. I personally find it rare to ever want strings to be concatenated like this, and when dealing with large `pd.DataFrame` objects, it can be prohibitively expensive to do so.

One way to avoid this issue is to be more explicit about the columns you would like to aggregate by selecting them after the `df.groupby("widget")` call:

```
df.groupby("widget")[["sales", "returns"]].agg("sum")
```

```
widget          sales     returns
```

```
Widget A        32          7
Widget B        36          12
```

Alternatively, you could reach for the pd.NamedAgg class we introduced back in the *Group by basics* recipe. Though more verbose, the use of pd.NamedAgg gives you the benefit of being able to rename the columns you would like to see in the output (i.e., instead of sales, you may want to see sales_total):

```
df.groupby("widget").agg(
    sales_total=pd.NamedAgg(column="sales", aggfunc="sum"),
    returns_total=pd.NamedAgg(column="returns", aggfunc="sum"),
)
```

```
widget          sales_total      returns_total
Widget A        32               7
Widget B        36               12
```

Another feature of pd.core.groupby.DataFrameGroupBy worth reviewing here is its ability to deal with multiple group arguments. By providing a list, you can expand your grouping to cover both widget and region:

```
df.groupby(["widget", "region"]).agg(
    sales_total=pd.NamedAgg("sales", "sum"),
    returns_total=pd.NamedAgg("returns", "sum"),
)
```

```
widget       region      sales_total      returns_total
Widget A     North       13               2
             South       19               5
Widget B     North       11               0
             South       25               12
```

With pd.core.groupby.DataFrameGroupBy.agg, there is no limitation on how many functions can be applied. For instance, if you want to see the sum, min, and mean of sales and returns within each widget and region, you could simply write the following:

```
df.groupby(["widget", "region"]).agg(
    sales_total=pd.NamedAgg("sales", "sum"),
    returns_total=pd.NamedAgg("returns", "sum"),
    sales_min=pd.NamedAgg("sales", "min"),
    returns_min=pd.NamedAgg("returns", "min"),
    sales_mean=pd.NamedAgg("sales", "mean"),
    returns_mean=pd.NamedAgg("returns", "mean"),
)
```

```
widget     region     sales_total     returns_total     sales_min     returns_min     sales_mean
    returns_mean
```

Widget A North 1.0	13	2	3	0	6.5
South 2.5	19	5	8	2	9.5
Widget B North 0.0	11	0	4	0	5.5
South 6.0	25	12	12	4	12.5

There's more...

While the built-in reduction functions and transformation functions that work out of the box with a group by are useful, there may still be times when you need to roll with your own custom function. This can be particularly useful when you find an algorithm to be good enough for what you are attempting in your local analysis, but when it may be difficult to generalize to all use cases.

A commonly requested function in pandas that is not provided out of the box with a group by is mode, even though there is a pd.Series.mode method. With pd.Series.mode, the type returned is always a pd.Series, regardless of whether there is only one value that appears most frequently:

```
pd.Series([0, 1, 1]).mode()
```

```
0    1
dtype: int64
```

This is true even if there are two or more elements that appear most frequently:

```
pd.Series([0, 1, 1, 2, 2]).mode()
```

```
0    1
1    2
dtype: int64
```

Given that there is a pd.Series.mode, why does pandas not offer a similar function when doing a group by? From a pandas developer perspective, the reason is simple; there is no single way to interpret what a group by should return.

Let's think through this in more detail with the following example, where group_a contains two values that appear with the same frequency (42 and 555), whereas group_b only contains the value 0:

```
df = pd.DataFrame([
    ["group_a", 42],
    ["group_a", 555],
    ["group_a", 42],
    ["group_a", 555],
    ["group_b", 0],
], columns=["group", "value"])

df
```

```
     group     value
0  group_a        42
1  group_a       555
2  group_a        42
3  group_a       555
4  group_b         0
```

The question we need to answer is *what should the mode return for group_a?* One possible solution would be to return a list (or any Python sequence) that holds both 42 and 555. The downside to this approach is that your returned dtype would be object, the pitfalls of which we covered back in *Chapter 3, Data Types*.

```
pd.Series([[42, 555], 0], index=pd.Index(["group_a", "group_b"], name="group"))
```

```
group
group_a     [42, 555]
group_b             0
dtype: object
```

A second expectation would be for pandas to just *choose one* of the values. Of course, this begs the question as to *how* pandas should make that decision – would the value 42 or 555 be more appropriate for group_a and how can that be determined in a general case?

A third expectation would be to return something where the label group_a appears twice in the resulting row index after aggregation. However, no other group by aggregations work this way, so we would be introducing new and potentially unexpected behavior by reducing to this:

```
pd.Series(
    [42, 555, 0],
    index=pd.Index(["group_a", "group_a", "group_b"], name="group")
)
```

```
group
group_a     42
group_a    555
group_b      0
dtype: int64
```

Rather than trying to solve for all of these expectations and codify it as part of the API, pandas leaves it entirely up to you how you would like to implement a mode function, as long as you adhere to the expectations that aggregations reduce to a single value per group. This eliminates the third expectation we just outlined as a possibility, at least until we talk about **Group by** apply later in this chapter.

To that end, if we wanted to roll with our own custom mode functions, they may end up looking something like:

```python
def scalar_or_list_mode(ser: pd.Series):
    result = ser.mode()
    if len(result) > 1:
        return result.tolist()
    elif len(result) == 1:
        return result.iloc[0]

    return pd.NA


def scalar_or_bust_mode(ser: pd.Series):
    result = ser.mode()
    if len(result) == 0:
        return pd.NA

    return result.iloc[0]
```

Since these are both aggregations, we can use them in the context of a `pd.core.groupby.DataFrameGroupBy.agg` operation:

```python
df.groupby("group").agg(
    scalar_or_list=pd.NamedAgg(column="value", aggfunc=scalar_or_list_mode),
    scalar_or_bust=pd.NamedAgg(column="value", aggfunc=scalar_or_bust_mode),
)
```

```
         scalar_or_list    scalar_or_bust
group
group_a     [42, 555]                  42
group_b             0                   0
```

Group by apply

During our discussion on algorithms and how to apply them back in *Chapter 5, Algorithms and How to Apply Them*, we came across the Apply function, which is both powerful and terrifying at the same time. An equivalent function for group by exists as `pd.core.groupby.DataFrameGroupBy.apply` with all of the same caveats. Generally, this function is overused, and you should opt for `pd.core.groupby.DataFrameGroupBy.agg` or `pd.core.groupby.DataFrameGroupBy.transform` instead. However, for the cases where you don't really want an *aggregation* or a *transformation*, but something in between, using apply is your only option.

Generally, `pd.core.groupby.DataFrameGroupBy.apply` should only be used as a last resort. It can produce sometimes ambiguous behavior and is rather prone to breakage across releases of pandas.

How to do it

In the *There's more...* section of the previous recipe, we mentioned how it is not possible to start with a pd.DataFrame of the following:

```
df = pd.DataFrame([
    ["group_a", 42],
    ["group_a", 555],
    ["group_a", 42],
    ["group_a", 555],
    ["group_b", 0],
], columns=["group", "value"])
df = df.convert_dtypes(dtype_backend="numpy_nullable")

df
```

```
   group    value
0  group_a     42
1  group_a    555
2  group_a     42
3  group_a    555
4  group_b      0
```

And to produce the following output, using a custom mode algorithm supplied to pd.core.groupby. DataFrameGroupBy.agg:

```
pd.Series(
    [42, 555, 0],
    index=pd.Index(["group_a", "group_a", "group_b"], name="group"),
    dtype=pd.Int64Dtype(),
)
```

```
group
group_a     42
group_a    555
group_b      0
dtype: Int64
```

The reason for this is straightforward; an aggregation expects you to reduce to a single value per group label. Repeating the label group_a twice in the output is a non-starter for an aggregation. Similarly, a transformation would expect you to produce a result that shares the same row index as the calling pd.DataFrame, which is not what we are after either.

`pd.core.groupby.DataFrameGroupBy.apply` is the in-between method that can get us closer to the desired result, which you can see in the following code. As a technical aside, the `include_groups=False` argument is passed to suppress any deprecation warnings about behavior in pandas 2.2. In subsequent versions, you may not need this:

```python
def mode_for_apply(df: pd.DataFrame):
    return df["value"].mode()

df.groupby("group").apply(mode_for_apply, include_groups=False)
```

```
group
group_a  0      42
         1     555
group_b  0       0
Name: value, dtype: Int64
```

It is important to note that we annotated the parameter of the `mode_for_apply` function as a `pd.DataFrame`. With aggregations and transformations, user-defined functions receive just a single `pd.Series` of data at a time, but with apply, you get an entire `pd.DataFrame`. For a more detailed look at what is going on, you can add `print` statements to the user-defined function:

```python
def mode_for_apply(df: pd.DataFrame):
    print(f"\nThe data passed to apply is:\n{df}")
    return df["value"].mode()

df.groupby("group").apply(mode_for_apply, include_groups=False)
```

```
The data passed to apply is:
   value
0     42
1    555
2     42
3    555

The data passed to apply is:
   value
4      0
group
group_a  0      42
         1     555
group_b  0       0
Name: value, dtype: Int64
```

Essentially, pd.core.groupby.DataFrameGroupBy.apply passes a pd.DataFrame of data to the user-defined function, excluding the column(s) that are used for grouping. From there, it will look at the return type of the user-defined function and try to infer the best possible output shape it can. In this particular instance, because our mode_for_apply function returns a pd.Series, pd.core.groupby. DataFrameGroupBy.apply has determined that the best output shape should have a pd.MultiIndex, where the first level of the index is the group value and the second level contains the row index from the pd.Series returned by the mode_for_apply function.

Where pd.core.groupby.DataFrameGroupBy.apply gets overused is in the fact that it can change its shape to look like an aggregation when it detects that the functions it applies reduce to a single value:

```python
def sum_values(df: pd.DataFrame):
    return df["value"].sum()


df.groupby("group").apply(sum_values, include_groups=False)
```

```
group
group_a     1194
group_b        0
dtype: int64
```

It is a trap to use it in this way, however. Even if it can infer a reasonable shape for some outputs, the rules for how it determines that are implementation details, for which you pay a performance penalty or run the risk of code breakage across pandas releases. If you know your functions will reduce to a single value, always opt for pd.core.groupby.DataFrameGroupBy.agg in lieu of pd.core.groupby. DataFrameGroupBy.apply, leaving the latter only for extreme use cases.

Window operations

Window operations allow you to calculate values over a sliding partition (or "window") of values. Commonly, these operations are used to calculate things like "rolling 90-day average," but they are flexible enough to extend to any algorithm of your choosing.

While not technically a group by operation, window operations are included here as they share a similar API and work with "groups" of data. The only difference to a group by call is that, instead of forming groups from unique value sets, a window operation creates its group by iterating over each value of a pandas object and looking at a particular number of preceding (and sometimes following) values.

How to do it

To get a feel for how window operations work, let's start with a simple pd.Series where each element is an increasing power of 2:

```
ser = pd.Series([0, 1, 2, 4, 8, 16], dtype=pd.Int64Dtype())
ser
```

```
0     0
1     1
2     2
3     4
4     8
5    16
dtype: Int64
```

The first type of window operation you will come across is the "rolling window," accessed via the pd.Series.rolling method. When calling this method, you need to tell pandas the desired size of your window *n*. The pandas library starts at each element and looks backward *n-1* records to form the "window":

```
ser.rolling(2).sum()
```

```
0     NaN
1     1.0
2     3.0
3     6.0
4    12.0
5    24.0
dtype: float64
```

You may notice that we started with a pd.Int64Dtype() but ended up with a float64 type after the rolling window operation. Unfortunately, the pandas window operations do not work well with the pandas extension system in at least version 2.2 (see issue #50449), so for the time being, we need to cast the result back into the proper data type:

```
ser.rolling(2).sum().astype(pd.Int64Dtype())
```

```
0    <NA>
1       1
2       3
3       6
4      12
5      24
dtype: Int64
```

So, what is going on here? Essentially, you can think of a rolling window operation as iterating through the pd.Series values. While doing so, it looks backward to try and collect enough values to fulfill the desired window size, which we have specified as 2.

After collecting two elements in each window, pandas will apply the specified aggregation function (in our case, summation). The result of that aggregation in each window is then used to piece back together the result:

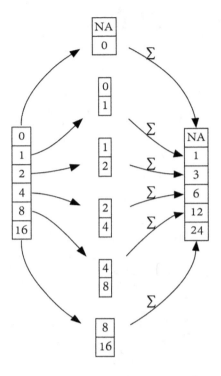

Figure 8.3: Rolling window with sum aggregation

In the case of our very first record, which cannot form a window with two elements, pandas returns a missing value. If you want the rolling calculation to just sum up as many elements as it can, even if the window size cannot be reached, you can pass an argument to min_periods= that dictates the minimum number of elements within each window required to perform the aggregation:

```
ser.rolling(2, min_periods=1).sum().astype(pd.Int64Dtype())
```

```
0     0
1     1
2     3
3     6
4     12
5     24
dtype: Int64
```

By default, rolling window operations look backward to try and fulfill your window size requirements. You can also "center" them instead so that pandas looks both forward and backward.

The effect of this is better seen with an odd window size. Note the difference when we expand our call so far with a window size of 3:

```
ser.rolling(3).sum().astype(pd.Int64Dtype())
```

```
0    <NA>
1    <NA>
2       3
3       7
4      14
5      28
dtype: Int64
```

Compared to the same call with an argument of center=True:

```
ser.rolling(3, center=True).sum().astype(pd.Int64Dtype())
```

```
0    <NA>
1       3
2       7
3      14
4      28
5    <NA>
dtype: Int64
```

Instead of looking at the current and preceding two values, usage of center=True tells pandas to take the current value, one prior, and one following to form a window.

Another type of window function is the "expanding window", which looks at all prior values encountered. The syntax for that is straightforward; simply replace your call to pd.Series.rolling with pd.Series.expanding and follow that up with your desired aggregation function. An expanding summation is similar to the pd.Series.cumsum method you have seen before, so for demonstration purposes, let's pick a different aggregation function, like mean:

```
ser.expanding().mean().astype(pd.Float64Dtype())
```

```
0         0.0
1         0.5
2         1.0
3        1.75
4         3.0
5    5.166667
dtype: Float64
```

Visually represented, an expanding window calculation looks as follows (for brevity, not all of the pd.Series elements are shown):

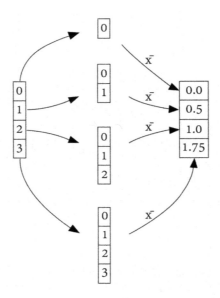

Figure 8.4: Expanding window with mean aggregation

There's more...

In *Chapter 10, General Usage and Performance Tips*, we will dive deeper into some of the very nice features pandas can offer when dealing with temporal data. Before we get there, it is worth noting that group by and rolling/expanding window functions work very naturally with such data, allowing you to concisely perform calculations like, "N day moving averages" "year-to-date X," "quarter-to-date X," etc.

To see how that works, let's take another look at the Nvidia stock performance dataset we started with back in *Chapter 6, Visualization*, originally as part of the *Calculating a trailing stop order price* recipe:

```
df = pd.read_csv(
    "data/NVDA.csv",
    usecols=["Date", "Close"],
    parse_dates=["Date"],
    dtype_backend="numpy_nullable",
).set_index("Date")
df
```

```
        Date       Close
2020-01-02    59.977501
2020-01-03    59.017502
2020-01-06    59.264999
2020-01-07    59.982498
2020-01-08    60.095001
...              ...
2023-12-22   488.299988
```

```
2023-12-26    492.790009
2023-12-27    494.170013
2023-12-28    495.220001
2023-12-29    495.220001
1006 rows × 1 columns
```

With rolling window functions, we can easily add 30, 60, and 90-day moving averages. A subsequent call to `pd.DataFrame.plot` also makes this easy to visualize:

```python
import matplotlib.pyplot as plt
plt.ion()

df.assign(
    ma30=df["Close"].rolling(30).mean().astype(pd.Float64Dtype()),
    ma60=df["Close"].rolling(60).mean().astype(pd.Float64Dtype()),
    ma90=df["Close"].rolling(90).mean().astype(pd.Float64Dtype()),
).plot()
```

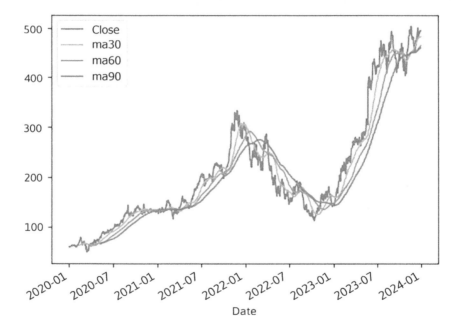

For "year-to-date" and "quarter-to-date" calculations, we can use a combination of group by and expanding window functions. For "year-to-date" min, max, and mean close values, we can start by forming a group by object to split our data into yearly buckets, and from there, we can make a call to `.expanding()`:

```python
df.groupby(pd.Grouper(freq="YS")).expanding().agg(
    ["min", "max", "mean"]
)
```

```
            Close
                      min          max          mean
Date        Date
2020-01-01  2020-01-02  59.977501   59.977501   59.977501
            2020-01-03  59.017502   59.977501   59.497501
            2020-01-06  59.017502   59.977501   59.420001
            2020-01-07  59.017502   59.982498   59.560625
            2020-01-08  59.017502   60.095001   59.667500
...         ...         ...          ...          ...
2023-01-01  2023-12-22  142.649994  504.089996   363.600610
            2023-12-26  142.649994  504.089996   364.123644
            2023-12-27  142.649994  504.089996   364.648024
            2023-12-28  142.649994  504.089996   365.172410
            2023-12-29  142.649994  504.089996   365.692600
1006 rows × 3 columns
```

The pd.Grouper(freq="YS") takes our row index, which contains datetimes, and groups them by the start of the year within which they fall. After the grouping, the call to .expanding() performs the min/max aggregations, only looking as far back as the start of each year. The effects of this are once again easier to see with a visualization:

```
df.groupby(pd.Grouper(freq="YS")).expanding().agg(
    ["min", "max", "mean"]
).droplevel(axis=1, level=0).reset_index(level=0, drop=True).plot()
```

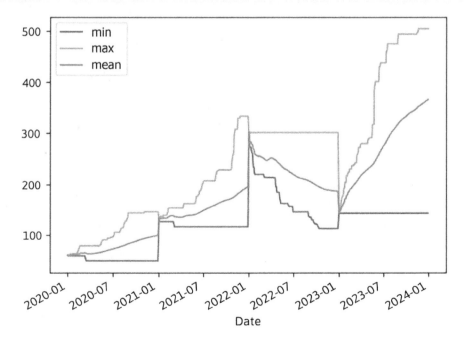

For a more granular view, you can calculate the expanding min/max close prices per quarter by changing the `freq=` argument from YS to QS in `pd.Grouper`:

```
df.groupby(pd.Grouper(freq="QS")).expanding().agg(
    ["min", "max", "mean"]
).reset_index(level=0, drop=True).plot()
```

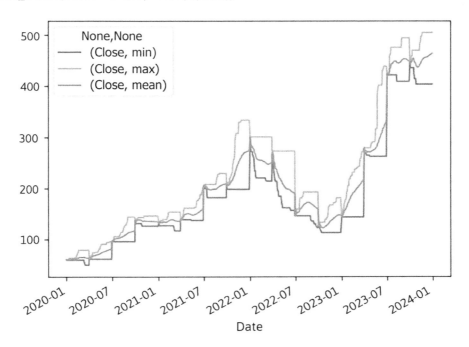

A `MS` `freq=` argument gets you down to the monthly level:

```
df.groupby(pd.Grouper(freq="MS")).expanding().agg(
    ["min", "max", "mean"]
).reset_index(level=0, drop=True).plot()
```

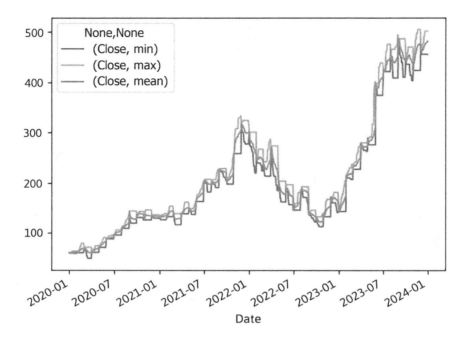

Selecting the highest rated movies by year

One of the most basic and common operations to perform during data analysis is to select rows containing the largest value of some column within a group. Applied to our movie dataset, this could mean finding the highest-rated film of each year or the highest-grossing film by content rating. To accomplish these tasks, we need to sort the groups as well as the column used to rank each member of the group, and then extract the highest member of each group.

In this recipe, we will find the highest-rated film of each year using a combination of `pd.DataFrame.sort_values` and `pd.DataFrame.drop_duplicates`.

How to do it

Start by reading in the movie dataset and slim it down to just the three columns we care about: `movie_title`, `title_year`, and `imdb_score`:

```
df = pd.read_csv(
    "data/movie.csv",
    usecols=["movie_title", "title_year", "imdb_score"],
    dtype_backend="numpy_nullable",
)

df
```

```
      movie_title                             title_year  imdb_score
0     Avatar                                  2009.0      7.9
1     Pirates of the Caribbean: At World's End  2007.0    7.1
2     Spectre                                 2015.0      6.8
3     The Dark Knight Rises                   2012.0      8.5
4     Star Wars: Episode VII - The Force Awakens  <NA>    7.1
...                                             ...         ...
4911  Signed Sealed Delivered                 2013.0      7.7
4912  The Following                           <NA>        7.5
4913  A Plague So Pleasant                    2013.0      6.3
4914  Shanghai Calling                        2012.0      6.3
4915  My Date with Drew                       2004.0      6.6
4916 rows × 3 columns
```

As you can see, the `title_year` column gets interpreted as a floating point value, but years should always be whole numbers. We could correct that by assigning the proper data type directly to our column:

```
df["title_year"] = df["title_year"].astype(pd.Int16Dtype())
df.head(3)
```

```
      movie_title                             title_year  imdb_score
0     Avatar                                  2009        7.9
1     Pirates of the Caribbean: At World's End  2007      7.1
2     Spectre                                 2015        6.8
```

Alternatively, we could have passed the desired data type as the `dtype=` argument in `pd.read_csv`:

```
df = pd.read_csv(
    "data/movie.csv",
    usecols=["movie_title", "title_year", "imdb_score"],
    dtype={"title_year": pd.Int16Dtype()},
    dtype_backend="numpy_nullable",
)
df.head(3)
```

```
      movie_title                             title_year  imdb_score
0     Avatar                                  2009        7.9
1     Pirates of the Caribbean: At World's End  2007      7.1
2     Spectre                                 2015        6.8
```

With our data cleansing out of the way, we can now turn our focus to answering the question of "what is the highest rated movie each year?". There are a few ways we can calculate this, but let's start with the approach you see most commonly.

When you perform a group by in pandas, the order in which rows appear in the original pd.DataFrame is respected as rows are bucketed into different groups. Knowing this, many users will answer this question by first sorting their dataset across title_year and imdb_score. After the sort, you can group by the title_year column, select just the movie_title column, and chain in a call to pd.DataFrameGroupBy. last to select the last value from each group:

```
df.sort_values(["title_year", "imdb_score"]).groupby(
    "title_year"
)[["movie_title"]].agg(top_rated_movie=pd.NamedAgg("movie_title", "last"))
```

```
title_year                                   top_rated_movie
1916        Intolerance: Love's Struggle Throughout the Ages
1920                            Over the Hill to the Poorhouse
1925                                          The Big Parade
1927                                               Metropolis
1929                                            Pandora's Box
...                                                       ...
2012                                        Django Unchained
2013            Batman: The Dark Knight Returns, Part 2
2014                                           Butterfly Girl
2015                                           Running Forever
2016                                    Kickboxer: Vengeance
91 rows × 1 columns
```

A slightly more succinct approach can be had if you use pd.DataFrameGroupBy.idxmax, which selects the row index value corresponding to the highest movie rating each year. This would require you to set the index to the movie_title up front:

```
df.set_index("movie_title").groupby("title_year").agg(
    top_rated_movie=pd.NamedAgg("imdb_score", "idxmax")
)
```

```
title_year                                   top_rated_movie
1916        Intolerance: Love's Struggle Throughout the Ages
1920                            Over the Hill to the Poorhouse
1925                                          The Big Parade
1927                                               Metropolis
1929                                            Pandora's Box
...                                                       ...
2012                                     The Dark Knight Rises
2013            Batman: The Dark Knight Returns, Part 2
2014                                  Queen of the Mountains
2015                                           Running Forever
```

2016	Kickboxer: Vengeance

91 rows × 1 columns

Our results appear mostly the same, although we can see that the two approaches disagreed on what the highest rated movie was in the years 2012 and 2014. A closer look at these titles reveals the root cause:

```
df[df["movie_title"].isin({
    "Django Unchained",
    "The Dark Knight Rises",
    "Butterfly Girl",
    "Queen of the Mountains",
})]
```

	movie_title	title_year	imdb_score
3	The Dark Knight Rises	2012	8.5
293	Django Unchained	2012	8.5
4369	Queen of the Mountains	2014	8.7
4804	Butterfly Girl	2014	8.7

In case of a tie, each method has its own way of choosing a value. Neither approach is right or wrong per se, but if you wanted finer control over that, you would have to reach for **Group by apply**.

Let's assume we wanted to aggregate the values so that when there is no tie, we get back a string, but in case of a tie we get a sequence of strings. To do this, you should define a function that accepts a pd.DataFrame. This pd.DataFrame will contain the values associated with each unique grouping column, which is title_year in our case.

Within the body of the function, you can figure out what the top movie rating is, find all movies with that rating, and return back either a single movie title (when there are no ties) or a set of movies (in case of a tie):

```
def top_rated(df: pd.DataFrame):
    top_rating = df["imdb_score"].max()
    top_rated = df[df["imdb_score"] == top_rating]["movie_title"].unique()

    if len(top_rated) == 1:
        return top_rated[0]
    else:
        return top_rated

df.groupby("title_year").apply(
    top_rated, include_groups=False
).to_frame().rename(columns={0: "top_rated_movie(s)"})
```

title_year	top_rated_movie(s)
1916	Intolerance: Love's Struggle Throughout the Ages

```
1920                                    Over the Hill to the Poorhouse
1925                                                  The Big Parade
1927                                                      Metropolis
1929                                                   Pandora's Box
...                                                              ...
2012                     [The Dark Knight Rises, Django Unchained]
2013                          Batman: The Dark Knight Returns, Part 2
2014                     [Queen of the Mountains, Butterfly Girl]
2015                                                 Running Forever
2016                                            Kickboxer: Vengeance
91 rows × 1 columns
```

Comparing the best hitter in baseball across years

In the *Finding the baseball players best at...* recipe back in *Chapter 6, Visualization*, we worked with a dataset that had already aggregated the performance of players from the years 2020-2023. However, comparing players based on their performance across multiple years is rather difficult. Even on a year-to-year basis, statistics that appear elite one year can be considered just "very good" in other years. The reasons for the variation in statistics across years can be debated, but likely come down to some combination of strategy, equipment, weather, and just pure statistical chance.

For this recipe, we are going to work with a more granular dataset that goes down to the game level. From there, we are going to aggregate the data up to a yearly summary, and from there calculate a common baseball statistic known as the *batting average*.

For those unfamiliar, a batting average is calculated by taking the number of *hits* a player produces (i.e., how many times they swung a bat at a baseball, and reached base as a result) as a percentage of their total *at bats* (i.e., how many times they came to bat, excluding *walks*).

So what constitutes a good batting average? As you will see, the answer to that question is a moving target, having shifted even within the past twenty years. In the early 2000s, a batting average between .260-.270 (i.e., getting a hit in 26%-27% of at bats) was considered middle of the road for professionals. Within recent years, that number has fallen somewhere in the range of .240-.250.

As such, to try and compare the *best hitters* from each year to one another, we cannot solely look at the batting average. A league-leading batting average of .325 in a year when the league itself averaged .240 is likely more impressive than a league-leading batting average of .330 in a year where the overall league averaged around .260.

How to do it

Once again, we are going to use data collected from retrosheet.org, with the following legal disclaimer:

For this recipe we are going to use box score summaries from every regular season game played in the years 2000-2023:

```
df = pd.read_parquet("data/mlb_batting_lines.parquet")

df
```

```
         year    game        starttime   ...   cs  gidp  int
0        2015    ANA201504100    7:12PM   ...    0     0    0
1        2015    ANA201504100    7:12PM   ...    0     0    0
2        2015    ANA201504100    7:12PM   ...    0     0    0
3        2015    ANA201504100    7:12PM   ...    0     0    0
4        2015    ANA201504100    7:12PM   ...    0     0    0
...       ...     ...             ...     ...   ...   ...  ...
1630995  2013    WAS201309222    7:06PM   ...    0     0    0
1630996  2013    WAS201309222    7:06PM   ...    0     0    0
1630997  2013    WAS201309222    7:06PM   ...    0     0    0
1630998  2013    WAS201309222    7:06PM   ...    0     0    0
1630999  2013    WAS201309222    7:06PM   ...    0     0    0
1631000 rows × 26 columns
```

A box score summarizes the performance of every player in a *game*. We could therefore single in on a particular game that was played in Baltimore on April 10, 2015, and see how batters performed:

```
bal = df[df["game"] == "BAL201504100"]
bal.head()
```

```
       year    game         starttime    ...   cs  gidp  int
2383   2015    BAL201504100    3:11PM     ...    0     0    0
2384   2015    BAL201504100    3:11PM     ...    0     0    0
2385   2015    BAL201504100    3:11PM     ...    0     0    0
2386   2015    BAL201504100    3:11PM     ...    0     0    0
2387   2015    BAL201504100    3:11PM     ...    0     0    0
5 rows × 26 columns
```

In that game alone we see a total of 75 at bats (*ab*), 29 hits (*h*) and two home runs (*hr*):

```
bal[["ab", "h", "hr"]].sum()
```

```
ab    75
h     29
hr     2
dtype: Int64
```

With a basic understanding of what a box score is and what it shows, let's turn our focus toward calculating the batting average every player produces each year. The individual player is notated in the id column of our dataset, and since we want to see the batting average over the course of an entire season, we can use the combination of year and id as our argument to pd.DataFrame.groupby. Afterward, we can apply a summation to the at bats (ab) and hits (h) columns:

```
df.groupby(["year", "id"]).agg(
    total_ab=pd.NamedAgg(column="ab", aggfunc="sum"),
    total_h=pd.NamedAgg(column="h", aggfunc="sum"),
)
```

```
year   id         total_ab   total_h
2000   abboj002      215        59
       abbok002      157        34
       abbop001        5         2
       abreb001      576       182
       acevj001        1         0
...    ...           ...       ...
2023   zavas001      175        30
       zerpa001        0         0
       zimmb002        0         0
       zunig001        0         0
       zunim001      124        22
31508 rows × 2 columns
```

To turn those totals into a batting average, we can chain in a division using pd.DataFrame.assign. After that, a call to pd.DataFrame.drop will let us solely focus on the batting average, dropping the total_ab and total_h columns we no longer need:

```
(
    df.groupby(["year", "id"]).agg(
        total_ab=pd.NamedAgg(column="ab", aggfunc="sum"),
        total_h=pd.NamedAgg(column="h", aggfunc="sum"))
    .assign(avg=lambda x: x["total_h"] / x["total_ab"])
    .drop(columns=["total_ab", "total_h"])
)
```

```
year   id         avg
2000   abboj002   0.274419
       abbok002   0.216561
       abbop001   0.400000
       abreb001   0.315972
       acevj001   0.000000
...    ...        ...
```

```
 2023   zavas001   0.171429
        zerpa001   NaN
        zimmb002   NaN
        zunig001   NaN
        zunim001   0.177419
31508 rows × 1 columns
```

Before we continue, we have to consider some data quality issues that may arise when calculating averages. Over the course of a baseball season, teams may use players who only appear in very niche situations, yielding a low number of plate appearances. In some instances, a batter may not even register an "at bat" for the season, so using that as a divisor has a chance of dividing by 0, which will produce NaN. In cases where a batter has a non-zero amount of at bats on the season, but still has relatively few, a small sample size can severely skew their batting average.

Major League Baseball has strict rules for determining how many plate appearances it takes for a batter to qualify for records within a given year. Without following the rule exactly, and without having to calculate plate appearances in our dataset, we can proxy this by setting a requirement of at least 400 at bats over the course of a season:

```python
(
    df.groupby(["year", "id"]).agg(
        total_ab=pd.NamedAgg(column="ab", aggfunc="sum"),
        total_h=pd.NamedAgg(column="h", aggfunc="sum"))
    .loc[lambda df: df["total_ab"] > 400]
    .assign(avg=lambda x: x["total_h"] / x["total_ab"])
    .drop(columns=["total_ab", "total_h"])
)
```

```
year   id         avg
2000   abreb001   0.315972
       alfoe001   0.323529
       alicl001   0.294444
       alomr001   0.309836
       aloum001   0.354626
...    ...        ...
2023   walkc002   0.257732
       walkj003   0.276190
       wittb002   0.276131
       yelic001   0.278182
       yoshm002   0.288641
4147 rows × 1 columns
```

We can summarize this further by finding the average and maximum batting_average per season, and we can even use pd.core.groupby.DataFrameGroupBy.idxmax to identify the player who achieved the best average:

```python
averages = (
    df.groupby(["year", "id"]).agg(
        total_ab=pd.NamedAgg(column="ab", aggfunc="sum"),
        total_h=pd.NamedAgg(column="h", aggfunc="sum"))
    .loc[lambda df: df["total_ab"] > 400]
    .assign(avg=lambda x: x["total_h"] / x["total_ab"])
    .drop(columns=["total_ab", "total_h"])
)

averages.groupby("year").agg(
    league_mean_avg=pd.NamedAgg(column="avg", aggfunc="mean"),
    league_max_avg=pd.NamedAgg(column="avg", aggfunc="max"),
    batting_champion=pd.NamedAgg(column="avg", aggfunc="idxmax"),
)
```

```
year   league_mean_avg   league_max_avg   batting_champion
2000   0.284512          0.372414         (2000, heltt001)
2001   0.277945          0.350101         (2001, walkl001)
2002   0.275713          0.369727         (2002, bondb001)
2003   0.279268          0.358714         (2003, pujoa001)
2004   0.281307          0.372159         (2004, suzui001)
2005   0.277350          0.335017         (2005, lee-d002)
2006   0.283609          0.347409         (2006, mauej001)
2007   0.281354          0.363025         (2007, ordom001)
2008   0.277991          0.364465         (2008, jonec004)
2009   0.278010          0.365201         (2009, mauej001)
2010   0.271227          0.359073         (2010, hamij003)
2011   0.269997          0.344406         (2011, cabrm001)
2012   0.269419          0.346405         (2012, cabrm002)
2013   0.268789          0.347748         (2013, cabrm001)
2014   0.267409          0.340909         (2014, altuj001)
2015   0.268417          0.337995         (2015, cabrm001)
2016   0.270181          0.347826         (2016, lemad001)
2017   0.268651          0.345763         (2017, altuj001)
2018   0.261824          0.346154         (2018, bettm001)
2019   0.269233          0.335341         (2019, andet001)
2021   0.262239          0.327731         (2021, turnt001)
2022   0.255169          0.326454         (2022, mcnej002)
2023   0.261457          0.353659         (2023, arral001)
```

As we can see, the mean batting average fluctuates each year, with those numbers having been higher back toward the year 2000. In the year 2005, the mean batting average was .277, with the best hitter (lee-d002, or Derrek Lee) having hit .335. The best hitter in 2019 (andet001, or Tim Anderson) also averaged .335, but the overall league was down around .269. Therefore, a strong argument could be made that Tim Anderson's 2019 season was more impressive than Derrek Lee's 2005 season, at least through the lens of batting average.

While taking the mean can be useful, it doesn't tell the full story of what goes on within a given season. We would probably like to get a better feel for the overall distribution of batting averages across each season, for which a visualization is in order. The violin plot we discovered back in the *Plotting movie ratings by decade with seaborn* recipe can help us understand this in more detail.

First let's set up our seaborn import, and have Matplotlib draw plots as soon as possible:

```python
import matplotlib.pyplot as plt
import seaborn as sns


plt.ion()
```

Next, we will want to make a few considerations for seaborn. Seaborn does not make use of `pd.MultiIndex`, so we are going to move our index values to columns with a call to `pd.DataFrame.reset_index`. Additionally, seaborn can easily misinterpret discrete *year* values like 2000, 2001, 2002, and so on for a continuous range, which we can solve by turning that column into a categorical data type.

The `pd.CategoricalDtype` we want to construct is also ideally ordered, so that pandas can ensure the year 2000 is followed by 2001, which is followed by 2002, and so on:

```python
sns_df = averages.reset_index()
years = sns_df["year"].unique()
cat = pd.CategoricalDtype(sorted(years), ordered=True)
sns_df["year"] = sns_df["year"].astype(cat)


sns_df
```

```
      year       id       avg
0     2000  abreb001  0.315972
1     2000  alfoe001  0.323529
2     2000  alicl001  0.294444
3     2000  alomr001  0.309836
4     2000  aloum001  0.354626
...    ...       ...       ...
4142  2023  walkc002  0.257732
4143  2023  walkj003  0.276190
4144  2023  wittb002  0.276131
4145  2023  yelic001  0.278182
4146  2023  yoshm002  0.288641
4147 rows × 3 columns
```

23 years of data on a single plot may take up a lot of space, so let's just look at the years 2000-2009 first:

```python
mask = (sns_df["year"] >= 2000) & (sns_df["year"] < 2010)
fig, ax = plt.subplots()
sns.violinplot(
    data=sns_df[mask],
    ax=ax,
    x="avg",
    y="year",
    order=sns_df.loc[mask, "year"].unique(),
)
ax.set_xlim(0.15, 0.4)
plt.show()
```

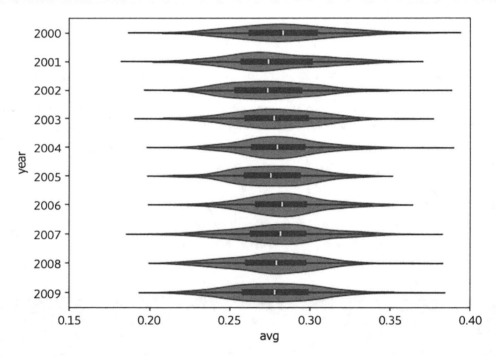

We intentionally made the call to `plt.subplots()` and used `ax.set_xlim(0.15, 0.4)` so that the x-axis would not change when plotting the remaining years:

```python
mask = sns_df["year"] >= 2010
fig, ax = plt.subplots()
sns.violinplot(
    data=sns_df[mask],
    ax=ax,
    x="avg",
```

```
    y="year",
    order=sns_df.loc[mask, "year"].unique(),
)
ax.set_xlim(0.15, 0.4)
plt.show()
```

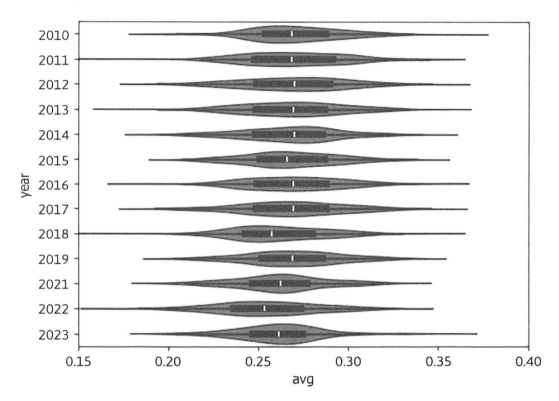

While some years show skew in the data (e.g., 2014 skewing right and 2018 skewing left), we can generally imagine the distribution of this data as an approximation of a normal distribution. Therefore, to try and better compare the peak performances across different years, we can use a technique whereby we *normalize* data within each season. Rather than thinking in terms of absolute batting averages like .250, we instead think of how far beyond the norm within a season a batter's performance is.

More specifically, we can use Z-score normalization, which would appear as follows when mathematically represented:

```
\begin{equation}
z=\frac{x - \mu} {\sigma}
\end{equation}
```

Here, μ is the mean and σ is the standard deviation.

Calculating this in pandas is rather trivial; all we need to do is define our custom `normalize` function and use that as an argument to `pd.core.groupby.DataFrameGroupBy.transform` to assign each combination of year and player their normalized batting average. Using that in subsequent group by operations allows us to better compare the peak performance each year across different years:

```python
def normalize(ser: pd.Series) -> pd.Series:
    return (ser - ser.mean()) / ser.std()


(
    averages.assign(
        normalized_avg=averages.groupby("year").transform(normalize)
    )
    .groupby("year").agg(
        league_mean_avg=pd.NamedAgg(column="avg", aggfunc="mean"),
        league_max_avg=pd.NamedAgg(column="avg", aggfunc="max"),
        batting_champion=pd.NamedAgg(column="avg", aggfunc="idxmax"),
        max_normalized_avg=pd.NamedAgg(column="normalized_avg", aggfunc="max"),
    )
    .sort_values(by="max_normalized_avg", ascending=False)
).head()
```

year	league_mean_avg	league_max_avg	batting_champion	max_normalized_avg
2023	0.261457	0.353659	(2023, arral001)	3.714121
2004	0.281307	0.372159	(2004, suzui001)	3.699129
2002	0.275713	0.369727	(2002, bondb001)	3.553521
2010	0.271227	0.359073	(2010, hamij003)	3.379203
2008	0.277991	0.364465	(2008, jonec004)	3.320429

According to this analysis, the 2023 season by Luis Arráez is the most impressive batting average performance since the year 2000. His `league_max_avg` achieved that year may appear as the lowest out of our top five, but so was the `league_mean_avg` in 2023.

As you can see from this recipe, effective use of pandas' Group By functionality allows you to more fairly evaluate records within different groups. Our example used professional baseball players within a season, but that same methodology could be extended to evaluate users within different age groups, products within different product lines, stocks within different sectors, and so on. Simply put, the possibilities for exploring your data with group by are endless!

Temporal Data Types and Algorithms

Properly working with temporal data (i.e., dates and times) may appear straightforward, but, the further you dive into it, the further you realize how surprisingly complex it is. Here are just a few issues that come to mind:

- Some users measure time in the span of years; others measure in nanoseconds
- Some users ignore timezones; others need to coordinate events around the world
- Not every country has multiple timezones, even if they are wide enough to have them (e.g., China)
- Not every country observes daylight saving time; those that do cannot agree on when
- In countries that observe daylight saving time, not every region participates (e.g., Arizona in the United States (US))
- Different operating systems and versions time differently (see also the Year 2038 problem at https://en.wikipedia.org/wiki/Year_2038_problem)

These problems are really just the tip of the iceberg, and, in spite of all of the potential data quality problems, temporal data is invaluable for purposes of monitoring, trend detection, and forecasting. Fortunately, pandas makes it so that you don't need to be an expert in dates and times to draw insights from your data. By using the features and abstractions pandas offers, you can very easily cleanse and interpolate your temporal data so that you can focus less on the "problems" of dates and times, and more on the insights that your data has to offer.

While we introduced some of the temporal types pandas has to offer back in *Chapter 3*, *Data Types*, in the section *Temporal types – datetime*, this chapter will start by focusing on things that pandas offers to augment the utility of those types. Beyond that, we will talk about the different ways you can cleanse and interpolate your temporal data, before finishing the chapter with a focus on practical applications.

We are going to cover the following recipes in this chapter:

- Timezone handling
- DateOffsets
- Datetime selection
- Resampling
- Aggregating weekly crime and traffic accidents
- Calculating year over year changes in crime by category
- Accurately measuring sensor-collected events with missing values

Timezone handling

By far, the most common mistakes with temporal data that I come across stem from a misunderstanding of timezones. On the East Coast of the US where I live, I've witnessed many users try to read what they think is a date of 2024-01-01 out of a database, yet ironically end up with a date of 2023-12-31 in their analysis. While that is only offset by a day, the effects of that misalignment can greatly skew summaries that group dates into weekly, monthly, quarterly, or yearly buckets.

For those that have been bitten by an issue like that before, you may have already come to realize that the source system you were communicating with probably did give you a timestamp of 2024-01-01 00:00:00, presumed to be at midnight UTC. Somewhere along the line, an analyst on the East Coast of the US where I live may have had that translated into their *local* time, which is either four hours offset from UTC during daylight saving time, or five hours offset during standard time. As a result, the timestamp ended up being viewed as 2023-12-31 20:00:00 or 2023-12-31 19:00:00 in EDT/EST, respectively, and the user may have inadvertently tried to convert that to a date.

To avoid these types of issues when working with temporal data, it is critical to understand when you are working with *timezone-aware* datetimes (i.e., those tied to a timezone like UTC or America/New_York), and *timezone-naive* objects, which have no timezone information attached to them. In this recipe, we will show you how to create and recognize both types of datetimes, while also diving deeper into the utilities pandas offers that let you convert between different timezones, and from timezone-aware to timezone-naive.

How to do it

Back in *Chapter 3*, *Data Types*, we learned how to create a pd.Series with datetime data. Let's take a closer look at that same example:

```
ser = pd.Series([
    "2024-01-01 00:00:00",
    "2024-01-02 00:00:01",
    "2024-01-03 00:00:02"
], dtype="datetime64[ns]")
ser
```

```
0    2024-01-01 00:00:00
1    2024-01-02 00:00:01
2    2024-01-03 00:00:02
dtype: datetime64[ns]
```

These timestamps represent events that occurred at or close to midnight on days ranging from January 1 through January 3, 2024. However, what these datetimes cannot tell us is *where* these events occurred; midnight in New York City happens at a different point in time than in Dubai, so it is tough to pinpoint an exact point in time that these events happened. Without that extra metadata, these datetimes are *timezone-naive*.

For programmatic confirmation that your datetimes are timezone-naive, you can use pd.Series. dt.tz, which will return None:

```
ser.dt.tz is None
```

```
True
```

With the pd.Series.dt.tz_localize method, we could assign an **Internet Assigned Numbers Authority (IANA)** timezone identifier to these datetimes to make them *timezone-aware*. For example, to specify that these events happened on the East Coast of the US, we could write:

```
ny_ser = ser.dt.tz_localize("America/New_York")
ny_ser
```

```
0    2024-01-01 00:00:00-05:00
1    2024-01-02 00:00:01-05:00
2    2024-01-03 00:00:02-05:00
dtype: datetime64[ns, America/New_York]
```

If you try to use pd.Series.dt.tz on this pd.Series, it will report back that you are working with a timezone of America/New_York:

```
ny_ser.dt.tz
```

```
<DstTzInfo 'America/New_York' LMT-1 day, 19:04:00 STD>
```

Now that our pd.Series is timezone-aware, the datetimes contained therein can be mapped to a point in time anywhere around the world. By using pd.Series.dt.tz_convert, you can easily translate these events into another timezone:

```
la_ser = ny_ser.dt.tz_convert("America/Los_Angeles")
la_ser
```

```
0    2023-12-31 21:00:00-08:00
1    2024-01-01 21:00:01-08:00
2    2024-01-02 21:00:02-08:00
dtype: datetime64[ns, America/Los_Angeles]
```

As a matter of practice, it is usually best to keep your datetimes attached to a timezone, which will mitigate the risk of being misinterpreted on a different date or at a different point in time. However, not all systems and databases that you may interact with will be able to retain this information, forcing you to drop it for interoperability. In case such a need arises, you could do this by passing None as an argument to pd.Series.dt.tz_localize:

```
la_ser.dt.tz_localize(None)
```

```
0    2023-12-31 21:00:00
1    2024-01-01 21:00:01
2    2024-01-02 21:00:02
dtype: datetime64[ns]
```

If you are forced to drop the timezone from your datetime, I would strongly recommend storing the timezone as a string in another column in your pd.DataFrame and database:

```
df = la_ser.to_frame().assign(
    datetime=la_ser.dt.tz_localize(None),
    timezone=str(la_ser.dt.tz),
).drop(columns=[0])

df
```

	datetime	timezone
0	2023-12-31 21:00:00	America/Los_Angeles
1	2024-01-01 21:00:01	America/Los_Angeles
2	2024-01-02 21:00:02	America/Los_Angeles

When roundtripping data like this, you can recreate the original pd.Series by applying the value from the timezone column to the data in the datetime column. For added safety, the following code sample uses the combination of pd.Series.drop_duplicates with pd.Series.squeeze to extract the single value of America/Los_Angeles from the timezone column before passing it to pd.Series.dt.tz_localize:

```
tz = df["timezone"].drop_duplicates().squeeze()
df["datetime"].dt.tz_localize(tz)
```

```
0    2023-12-31 21:00:00-08:00
1    2024-01-01 21:00:01-08:00
2    2024-01-02 21:00:02-08:00
Name: datetime, dtype: datetime64[ns, America/Los_Angeles]
```

DateOffsets

In the *Temporal types – Timedelta* recipe back in *Chapter 3, Data Types*, we introduced the `pd.Timedelta` type and mentioned how it could be used to shift datetimes by a finite duration, like 10 seconds or 5 days. However, a `pd.Timedelta` cannot be used to offset a date or datetime by say *one month* because a month does not always represent the same duration of time. In the Gregorian calendar, months can range in duration from 28–31 days. The month of February is usually 28 days but extends to 29 days for every year that is divisible by 4, unless the year is divisible by 100 but not by 400.

Thinking about these issues all of the time would be rather tedious. Fortunately, pandas takes care of all of the mundane details and just lets you shift dates according to a calendar through the use of the `pd.DateOffset` object, which we will explore in this recipe.

How to do it

To build a foundational knowledge of how this works, let's start with a very simple `pd.Series` containing the first few days of 2024:

```
ser = pd.Series([
    "2024-01-01",
    "2024-01-02",
    "2024-01-03",
], dtype="datetime64[ns]")
ser
```

```
0    2024-01-01
1    2024-01-02
2    2024-01-03
dtype: datetime64[ns]
```

Shifting these dates by one month would typically mean keeping the same day of the month, but just placing the dates in February instead of January. With `pd.DateOffset`, you can pass in an argument to `months=` that dictates the number of months you want to move the dates by; so, let's see how it looks with an argument of 1:

```
ser + pd.DateOffset(months=1)
```

```
0    2024-02-01
1    2024-02-02
2    2024-02-03
dtype: datetime64[ns]
```

Shifting by two months would mean moving these dates from January to March. We shouldn't really care that there were 31 days in January but 29 in February 2024; the `pd.DateOffset` takes care of this for us:

```
ser + pd.DateOffset(months=2)
```

```
0    2024-03-01
1    2024-03-02
2    2024-03-03
dtype: datetime64[ns]
```

For dates that wouldn't exist (e.g., trying to shift January 30 to February 30), pd.DateOffset will try and match to the closest date that does exist within the target month:

```
pd.Series([
    "2024-01-29",
    "2024-01-30",
    "2024-01-31",
], dtype="datetime64[ns]") + pd.DateOffset(months=1)
```

```
0    2024-02-29
1    2024-02-29
2    2024-02-29
dtype: datetime64[ns]
```

You can also step backward through the calendar with a negative argument to months=:

```
ser + pd.DateOffset(months=-1)
```

```
0    2023-12-01
1    2023-12-02
2    2023-12-03
dtype: datetime64[ns]
```

The pd.DateOffset is flexible enough to accept more than just one keyword argument at a time. For instance, if you wanted to offset your dates by one month, two days, three hours, four minutes, and five seconds, you could do that all in one expression:

```
ser + pd.DateOffset(months=1, days=2, hours=3, minutes=4, seconds=5)
```

```
0    2024-02-03 03:04:05
1    2024-02-04 03:04:05
2    2024-02-05 03:04:05
dtype: datetime64[ns]
```

Alongside the pd.DateOffset class, pandas offers you the ability to shift dates to the beginning or the end of a period with various classes exposed in the pd.offsets module. For instance, if you want to shift your dates to the end of the month, you can use pd.offsets.MonthEnd:

```
ser + pd.offsets.MonthEnd()
```

```
0    2024-01-31
1    2024-01-31
```

```
2   2024-01-31
dtype: datetime64[ns]
```

`pd.offsets.MonthBegin` will move the dates to the beginning of the next month:

```
ser + pd.offsets.MonthBegin()
```

```
0   2024-02-01
1   2024-02-01
2   2024-02-01
dtype: datetime64[ns]
```

`pd.offsets.SemiMonthBegin`, `pd.offsets.SemiMonthEnd`, `pd.offsets.QuarterBegin`, `pd.offsets.QuarterEnd`, `pd.offsets.YearBegin`, and `pd.offsets.YearEnd` all offer similar behavior to shift your dates to the beginning or end of different periods.

There's more...

The `pd.DateOffset`, by default, works against the Gregorian calendar, but different subclasses of this can provide more customized functionality.

One of the most used subclasses is the `pd.offsets.BusinessDay`, which, by default, only counts the standard "business days" of Monday through Friday when offsetting dates. To see how this works, let's consider the day of the week each of our dates in `ser` fall on:

```
ser.dt.day_name()
```

```
0       Monday
1      Tuesday
2    Wednesday
dtype: object
```

Now, let's see what happens when we add three business days to our dates:

```
bd_ser = ser + pd.offsets.BusinessDay(n=3)
bd_ser
```

```
0   2024-01-04
1   2024-01-05
2   2024-01-08
dtype: datetime64[ns]
```

We can use the same `pd.Series.dt.day_name` method to check the new days of the week that these dates fall on:

```
bd_ser.dt.day_name()
```

```
0     Thursday
1       Friday
```

```
2        Monday
dtype: object
```

After having added three business days, our dates that started on Monday and Tuesday ended up falling on the Thursday and Friday of the same week, respectively. The Wednesday date we started with was pushed to the Monday of the following week, as neither Saturday nor Sunday qualifies as a business day.

If you work with a business that has different business days from Monday to Friday, you could use the `pd.offsets.CustomBusinessDay` to set up your own rules for how offsetting should work. The argument to `weekmask=` will dictate the days of the week that are considered business days:

```
ser + pd.offsets.CustomBusinessDay(
    n=3,
    weekmask="Mon Tue Wed Thu",
)
```

```
0    2024-01-04
1    2024-01-08
2    2024-01-09
dtype: datetime64[ns]
```

You can even add a `holidays=` argument to account for days when your business may be closed:

```
ser + pd.offsets.CustomBusinessDay(
    n=3,
    weekmask="Mon Tue Wed Thu",
    holidays=["2024-01-04"],
)
```

```
0    2024-01-08
1    2024-01-09
2    2024-01-10
dtype: datetime64[ns]
```

For the Gregorian calendar, we have already seen `pd.offsets.MonthEnd` and `pd.offsets.MonthBegin` classes that help you move dates to the beginning or end of a month, respectively. Similar classes exist for you to use when attempting to shift dates toward the beginning or end of business months:

```
ser + pd.offsets.BusinessMonthEnd()
```

```
0    2024-01-31
1    2024-01-31
2    2024-01-31
dtype: datetime64[ns]
```

Datetime selection

Back in *Chapter 2*, *Selection and Assignment*, we discussed the many robust ways that pandas allows you to select data from a pd.Series or pd.DataFrame by interacting with their associated row pd.Index. If you happen to create a pd.Index using datetime data, it ends up being represented as a special subclass called a pd.DatetimeIndex. This subclass overrides some functionality of the pd.Index.loc method to give you more flexible selection options tailored to temporal data.

How to do it

pd.date_range is a convenient function that helps you quickly generate a pd.DatetimeIndex. One of the ways to use this function is to specify a starting date with the start= parameter, specify a step frequency with the freq= parameter, and specify the desired length of your pd.DatetimeIndex with the periods= argument.

For instance, to generate a pd.DatetimeIndex that starts on December 27, 2023, and provides 5 days in total with 10 days between each record, you would write:

```
pd.date_range(start="2023-12-27", freq="10D", periods=5)
```

```
DatetimeIndex(['2023-12-27', '2024-01-06', '2024-01-16', '2024-01-26',
               '2024-02-05'],
              dtype='datetime64[ns]', freq='10D')
```

A frequency string of "2W" will generate dates spaced two weeks apart. If the start= parameter is a Sunday, the dates will begin from that date exactly; otherwise, the next Sunday begins the sequence:

```
pd.date_range(start="2023-12-27", freq="2W", periods=5)
```

```
DatetimeIndex(['2023-12-31', '2024-01-14', '2024-01-28', '2024-02-11',
               '2024-02-25'],
              dtype='datetime64[ns]', freq='2W-SUN')
```

You could even control the day of the week being used to anchor the dates by appending a suffix like "-WED", which will generate dates on Wednesday instead of Sunday:

```
pd.date_range(start="2023-12-27", freq="2W-WED", periods=5)
```

```
DatetimeIndex(['2023-12-27', '2024-01-10', '2024-01-24', '2024-02-07',
               '2024-02-21'],
              dtype='datetime64[ns]', freq='2W-WED')
```

A freq= argument of "WOM-3THU" will give you the third Thursday of every month:

```
pd.date_range(start="2023-12-27", freq="WOM-3THU", periods=5)
```

```
DatetimeIndex(['2024-01-18', '2024-02-15', '2024-03-21', '2024-04-18',
               '2024-05-16'],
              dtype='datetime64[ns]', freq='WOM-3THU')
```

The first and fifteenth day of each month can be generated with an argument of `"SMS"`:

```
pd.date_range(start="2023-12-27", freq="SMS", periods=5)
```

```
DatetimeIndex(['2024-01-01', '2024-01-15', '2024-02-01', '2024-02-15',
               '2024-03-01'],
              dtype='datetime64[ns]', freq='SMS-15')
```

As you can see, there are countless frequency strings that can be used to describe what pandas refers to as **date offsets**. For a more complete listing, be sure to reference the pandas documentation at https://pandas.pydata.org/pandas-docs/stable/user_guide/timeseries.html#dateoffset-objects.

Each element of the `pd.DatetimeIndex` is actually a `pd.Timestamp`. When using this for selection from a `pd.Series` or `pd.DataFrame`, users may at first be tempted to write something like the following to select all records up to and including a date like 2024-01-18:

```
index = pd.date_range(start="2023-12-27", freq="10D", periods=20)
ser = pd.Series(range(20), index=index)
ser.loc[:pd.Timestamp("2024-01-18")]
```

```
2023-12-27    0
2024-01-06    1
2024-01-16    2
Freq: 10D, dtype: int64
```

Similarly, users may be tempted to write the following to select a range of dates:

```
ser.loc[pd.Timestamp("2024-01-06"):pd.Timestamp("2024-01-18")]
```

```
2024-01-06    1
2024-01-16    2
Freq: 10D, dtype: int64
```

However, these methods of selecting from a `pd.DatetimeIndex` are rather verbose. For convenience, pandas lets you pass in strings to represent the desired dates, instead of `pd.Timestamp` instances:

```
ser.loc["2024-01-06":"2024-01-18"]
```

```
2024-01-06    1
2024-01-16    2
Freq: 10D, dtype: int64
```

You also are not required to specify the entire date in YYYY-MM-DD format. For instance, if you wanted to select all of the dates that fall in February 2024, you could just pass the string `2024-02` to your `pd.Series.loc` call:

```
ser.loc["2024-02"]
```

```
2024-02-05     4
2024-02-15     5
2024-02-25     6
Freq: 10D, dtype: int64
```

Slicing will be intelligent enough to recognize this pattern, making it easy to select all of the records in both February and March:

```
ser.loc["2024-02":"2024-03"]
```

```
2024-02-05     4
2024-02-15     5
2024-02-25     6
2024-03-06     7
2024-03-16     8
2024-03-26     9
Freq: 10D, dtype: int64
```

You can take this abstraction a step further and select an entire year:

```
ser.loc["2024"].head()
```

```
2024-01-06     1
2024-01-16     2
2024-01-26     3
2024-02-05     4
2024-02-15     5
Freq: 10D, dtype: int64
```

There's more...

A pd.DatetimeIndex can also be associated with a timezone by providing a tz= argument:

```
index = pd.date_range(start="2023-12-27", freq="12h", periods=6, tz="US/
Eastern")
ser = pd.Series(range(6), index=index)
ser
```

```
2023-12-27 00:00:00-05:00     0
2023-12-27 12:00:00-05:00     1
2023-12-28 00:00:00-05:00     2
2023-12-28 12:00:00-05:00     3
2023-12-29 00:00:00-05:00     4
2023-12-29 12:00:00-05:00     5
Freq: 12h, dtype: int64
```

When using strings to select from a timezone-aware `pd.DatetimeIndex`, be aware that pandas will implicitly convert your string argument into the timezone of the `pd.DatetimeIndex`. For instance, the following code will only select one element from our data:

```
ser.loc[:"2023-12-27 11:59:59"]
```

```
2023-12-27 00:00:00-05:00     0
Freq: 12h, dtype: int64
```

Whereas the following code will correctly select both elements:

```
ser.loc[:"2023-12-27 12:00:00"]
```

```
2023-12-27 00:00:00-05:00     0
2023-12-27 12:00:00-05:00     1
Freq: 12h, dtype: int64
```

These both work in spite of the fact that our dates are five hours offset from UTC, and our string makes no indication of the expected timezone. In this way, pandas makes it very easy to express selection from a `pd.DatetimeIndex`, whether it is timezone-aware or timezone-naive.

Resampling

Back in *Chapter 8, Group By*, we went in-depth into the group by functionality that pandas has to offer. A Group By allows you to *split* your data based on unique value combinations in your dataset, *apply* an algorithm to those splits, and combine the results back together.

A *resample* is very similar to a Group By, with the only difference happening during the *split* phase. Instead of generating groups from unique value combinations, a resample lets you take datetimes and group them into increments like *every 5 seconds* or *every 10 minutes*.

How to do it

Let's once again reach for the `pd.date_range` function we were introduced to back in the *Datetime selection* recipe, but this time, we are going to generate a `pd.DatetimeIndex` with a frequency of seconds instead of days:

```
index = pd.date_range(start="2024-01-01", periods=10, freq="s")
ser = pd.Series(range(10), index=index, dtype=pd.Int64Dtype())
ser
```

```
2024-01-01 00:00:00     0
2024-01-01 00:00:01     1
2024-01-01 00:00:02     2
2024-01-01 00:00:03     3
2024-01-01 00:00:04     4
2024-01-01 00:00:05     5
2024-01-01 00:00:06     6
```

```
2024-01-01 00:00:07    7
2024-01-01 00:00:08    8
2024-01-01 00:00:09    9
Freq: s, dtype: Int64
```

If viewing this data every second was deemed too granular, `pd.Series.resample` can be used to *downsample* the data into a different increment, like *every 3 seconds*. Resampling also requires the use of an aggregation function to dictate what happens to all records that fall within each increment; for simplicity, we can start with summation:

```
ser.resample("3s").sum()
```

```
2024-01-01 00:00:00     3
2024-01-01 00:00:03    12
2024-01-01 00:00:06    21
2024-01-01 00:00:09     9
Freq: 3s, dtype: Int64
```

In this particular case, `resample` creates buckets using the ranges of `[00:00:00-00:00:03)`, `[00:00:03-00:00:06)`, `[00:00:06-00:00:09)`, and `[00:00:09-00:00:12)`. For each of those intervals, the left square bracket indicates that the interval is closed on the left side (i.e., it includes those values). By contrast, the right parentheses indicate an open interval that does not include the value.

Technically speaking, all of these intervals created by the resample with a frequency of `"3s"` are "left-closed" by default, but the `closed=` argument can be used to change that behavior, effectively producing intervals with the values of `(23:59:57-00:00:00]`, `(00:00:00-00:00:03]`, `(00:00:03-00:00:06]`, and `(00:00:06-00:00:09]`:

```
ser.resample("3s", closed="right").sum()
```

```
2023-12-31 23:59:57     0
2024-01-01 00:00:00     6
2024-01-01 00:00:03    15
2024-01-01 00:00:06    24
Freq: 3s, dtype: Int64
```

With the frequency of `"3s"`, the left value of the interval is used as the value in the resulting row index. That behavior can also be changed through the use of the `label=` argument:

```
ser.resample("3s", closed="right", label="right").sum()
```

```
2024-01-01 00:00:00     0
2024-01-01 00:00:03     6
2024-01-01 00:00:06    15
2024-01-01 00:00:09    24
Freq: 3s, dtype: Int64
```

One last caveat you may want to be aware of is that the default values for the `closed=` and `label=` arguments depend upon the frequency that you have chosen. Our frequency of `"3s"` creates left-closed intervals, and uses the left interval value in the row index. However, if we had chosen a frequency that is oriented toward the end of a period, like `ME` or `YE` (month-end and year-end, respectively), pandas will instead produce right-closed intervals and use the right label:

```
ser.resample("ME").sum()
```

```
2024-01-31    45
Freq: ME, dtype: Int64
```

While we are on the topic of downsampling, let's take a look at a different frequency, like days (`"D"`). At this level, `pd.Series.resample` can be a convenient way to aggregate daily events into weekly buckets. To see how this works, let's just look at the first 10 days of 2024:

```
index = pd.date_range(start="2024-01-01", freq="D", periods=10)
ser = pd.Series(range(10), index=index, dtype=pd.Int64Dtype())
ser
```

```
2024-01-01    0
2024-01-02    1
2024-01-03    2
2024-01-04    3
2024-01-05    4
2024-01-06    5
2024-01-07    6
2024-01-08    7
2024-01-09    8
2024-01-10    9
Freq: D, dtype: Int64
```

Without looking up which day of the week each of these falls on, we can use `pd.DatetimeIndex.dt.day_name()` to ground ourselves:

```
ser.index.day_name()
```

```
Index(['Monday', 'Tuesday', 'Wednesday', 'Thursday', 'Friday', 'Saturday',
       'Sunday', 'Monday', 'Tuesday', 'Wednesday'],
    dtype='object')
```

By default, resampling into weekly buckets will create periods that *end* on a Sunday:

```
ser.resample("W").sum()
```

```
2024-01-07    21
2024-01-14    24
Freq: W-SUN, dtype: Int64
```

You are free, however, to pick any day of the week for your period to end on. In the US, considering Saturday to be the end of the week is arguably more common than Sunday:

```
ser.resample("W-SAT").sum()
```

```
2024-01-06    15
2024-01-13    30
Freq: W-SAT, dtype: Int64
```

Though, you can pick any day of the week:

```
ser.resample("W-WED").sum()
```

```
2024-01-03     3
2024-01-10    42
Freq: W-WED, dtype: Int64
```

Now that we have covered the topic of *downsampling* (i.e., going from a more granular to a less granular frequency), let's take a look at going in the opposite direction with the process of *upsampling*. Our data shows events that happen every day, but what if we wanted to create a time series that measured events every 12 hours?

Fortunately, the API to achieve this is not all that different. You can still use pd.Series.resample to start, but will subsequently want to chain in a call to pandas.core.resample.Resampler.asfreq:

```
ser.resample("12h").asfreq().iloc[:5]
```

```
2024-01-01 00:00:00        0
2024-01-01 12:00:00     <NA>
2024-01-02 00:00:00        1
2024-01-02 12:00:00     <NA>
2024-01-03 00:00:00        2
Freq: 12h, dtype: Int64
```

Intervals generated during the *upsample*, which have no associated activity, are assigned a missing value. Left alone, there is likely not a ton of value to upsampling like this. However, pandas offers a few ways to fill in this missing data.

The first approach to handle missing data may be to forward fill or backward fill values, so that missing values are just replaced with whatever record came preceding or following, respectively.

A forward fill will generate values of [0, 0, 1, 1, 2, 2, ...]:

```
ser.resample("12h").asfreq().ffill().iloc[:6]
```

```
2024-01-01 00:00:00    0
2024-01-01 12:00:00    0
2024-01-02 00:00:00    1
2024-01-02 12:00:00    1
```

```
2024-01-03 00:00:00    2
2024-01-03 12:00:00    2
Freq: 12h, dtype: Int64
```

Whereas a backward fill yields [0, 1, 1, 2, 2, 3, ...]:

```
ser.resample("12h").asfreq().bfill().iloc[:6]
```

```
2024-01-01 00:00:00    0
2024-01-01 12:00:00    1
2024-01-02 00:00:00    1
2024-01-02 12:00:00    2
2024-01-03 00:00:00    2
2024-01-03 12:00:00    3
Freq: 12h, dtype: Int64
```

An arguably more robust solution can be had in the form of *interpolation*, where the values preceding and following a missing value can be used to mathematically guess the missing value. The default interpolation will be *linear*, essentially taking the average of the value before and after each missing value:

```
ser.resample("12h").asfreq().interpolate().iloc[:6]
```

```
2024-01-01 00:00:00    0.0
2024-01-01 12:00:00    0.5
2024-01-02 00:00:00    1.0
2024-01-02 12:00:00    1.5
2024-01-03 00:00:00    2.0
2024-01-03 12:00:00    2.5
Freq: 12h, dtype: Float64
```

There's more...

In the introduction to this recipe, we mentioned that a resample was similar to a Group By. In fact, you could rewrite a resample using pd.DataFrame.groupby with a pd.Grouper argument.

Let's once again look at a pd.Series with 10 records occurring every second:

```
index = pd.date_range(start="2024-01-01", periods=10, freq="s")
ser = pd.Series(range(10), index=index, dtype=pd.Int64Dtype())
ser
```

```
2024-01-01 00:00:00    0
2024-01-01 00:00:01    1
2024-01-01 00:00:02    2
2024-01-01 00:00:03    3
2024-01-01 00:00:04    4
2024-01-01 00:00:05    5
```

```
2024-01-01 00:00:06    6
2024-01-01 00:00:07    7
2024-01-01 00:00:08    8
2024-01-01 00:00:09    9
Freq: s, dtype: Int64
```

A resample into three-second increments looks as follows:

```
ser.resample("3s").sum()
```

```
2024-01-01 00:00:00     3
2024-01-01 00:00:03    12
2024-01-01 00:00:06    21
2024-01-01 00:00:09     9
Freq: 3s, dtype: Int64
```

This would be rewritten to get the same result by passing in "3s" to the freq= argument of a pd.Grouper:

```
ser.groupby(pd.Grouper(freq="3s")).sum()
```

```
2024-01-01 00:00:00     3
2024-01-01 00:00:03    12
2024-01-01 00:00:06    21
2024-01-01 00:00:09     9
Freq: 3s, dtype: Int64
```

There is no requirement that you use pd.DataFrame.resample, and, in fact, you will find that the pd.Grouper approach works better when you must also group by non-datetime values. We will see this in action in the *Calculating year-over-year changes in crime by category* recipe later in this chapter.

Aggregating weekly crime and traffic accidents

So far in this chapter, we have taken a basic tour of pandas' offerings for dealing with temporal data. Starting with small sample datasets has made it easy to visually inspect the output of our operations, but we are now at the point where we can start focusing on applications to "real world" datasets.

The Denver crime dataset is huge, with over 460,000 rows each marked with a datetime of when the crime was reported. As you will see in this recipe, we can use pandas to easily resample these events and ask questions like *How many crimes were reported in a given week?*.

How to do it

To start, let's read in the crime dataset, setting our index as the REPORTED_DATE. This dataset was saved using pandas extension types, so there is no need to specify the dtype_backend= argument:

```
df = pd.read_parquet(
    "data/crime.parquet",
).set_index("REPORTED_DATE")
```

```
df.head()
```

REPORTED_DATE	OFFENSE_TYPE_ID	OFFENSE_CATEGORY_ID
2014-06-29 02:01:00	traffic-accident-dui-duid	traffic-accident
2014-06-29 01:54:00	vehicular-eluding-no-chase	all-other-crimes
2014-06-29 02:00:00	disturbing-the-peace	public-disorder
2014-06-29 02:18:00	curfew	public-disorder
2014-06-29 04:17:00	aggravated-assault	aggravated-assault

GEO_LON	NEIGHBORHOOD_ID	IS_CRIME	IS_TRAFFIC
-105.000149	cbd	0	1
-105.020719	ath-mar-park	1	0
-105.001552	sunny-side	1	0
-105.018557	college-view-south-platte	1	0

5 rows × 7 columns

To count the number of crimes per week, we need to form a group for each week, which we know we can do with `pd.DataFrame.resample`. Chaining a call to the `.size` method will count the number of crimes within each week for us:

```
df.resample("W").size()
```

```
REPORTED_DATE
2012-01-08      877
2012-01-15     1071
2012-01-22      991
2012-01-29      988
2012-02-05      888
                ...
2017-09-03     1956
2017-09-10     1733
2017-09-17     1976
2017-09-24     1839
2017-10-01     1059
Freq: W-SUN, Length: 300, dtype: int64
```

We now have the weekly crime count as a `pd.Series` with the new index incrementing one week at a time. There are a few things that happen by default that are very important to understand. Sunday is chosen as the last day of the week and is also the date used to label each element in the resulting `pd.Series`. For instance, the first index value, January 8, 2012, is a Sunday. There were 877 crimes committed during that week ending on the 8th. The week of Monday, January 9, to Sunday, January 15, recorded 1,071 crimes. Let's do some sanity checks and ensure that our resampling is doing this:

```
len(df.sort_index().loc[:'2012-01-08'])
```

```
877
```

```
len(df.sort_index().loc['2012-01-09':'2012-01-15'])
```

```
1071
```

To get an overall understanding of the trend, it would be helpful to create a plot from our resampled data:

```
import matplotlib.pyplot as plt
plt.ion()
df.resample("W").size().plot(title="All Denver Crimes")
```

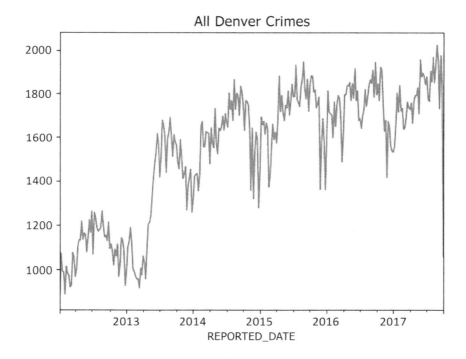

The Denver crime dataset has all crime and traffic accidents together in one table and separates them through the binary columns IS_CRIME and IS_TRAFFIC. Using pd.DataFrame.resample, we can select just these two columns and summarize them over a given period. For a quarterly summary, you would write:

```
df.resample("QS")[["IS_CRIME", "IS_TRAFFIC"]].sum().head()
```

	IS_CRIME	IS_TRAFFIC
REPORTED_DATE		
2012-01-01	7882	4726
2012-04-01	9641	5255
2012-07-01	10566	5003

| 2012-10-01 | 9197 | 4802 |
| 2013-01-01 | 8730 | 4442 |

Once again, a line plot to understand the trend may be more helpful:

```
df.resample("QS")[["IS_CRIME", "IS_TRAFFIC"]].sum().plot(
    color=["black", "lightgrey"],
    title="Denver Crime and Traffic Accidents"
)
```

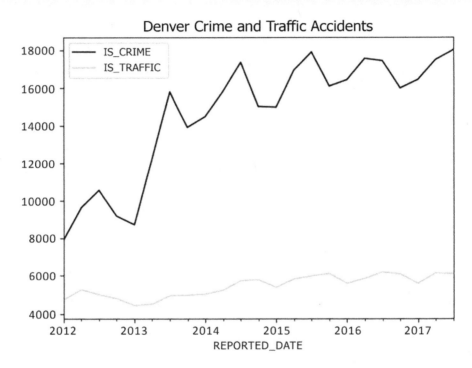

Calculating year-over-year changes in crime by category

Often, users want to know *How much did this change year over year?* or *...quarter over quarter?*. In spite of the frequency with which these questions are asked, writing algorithms to try and answer them can be rather complex and time-intensive. Fortunately, pandas gives you much of this functionality out of the box, trivializing much of the effort.

To try and make things more complicated, in this recipe, we are going to ask the question of *how much did it change by category?* Adding *by category* into the equation will prevent us from directly using pd.DataFrame.resample, but as you will see, pandas can still very easily help you answer these detailed types of questions.

How to do it

Let's read in the crime dataset, but this time, we are not going to set the REPORTED_DATE as our index:

```
df = pd.read_parquet(
    "data/crime.parquet",
)
df.head()
```

```
      OFFENSE_TYPE_ID    OFFENSE_CATEGORY_ID    REPORTED_DATE   ...   NEIGHBORHOOD_ID
IS_CRIME    IS_TRAFFIC
0    traffic-accident-dui-duid    traffic-accident    2014-06-29 02:01:00   ...
cbd    0    1
1    vehicular-eluding-no-chase    all-other-crimes    2014-06-29 01:54:00   ...
east-colfax    1    0
2    disturbing-the-peace    public-disorder    2014-06-29 02:00:00   ...    athmar-
park    1    0
3    curfew    public-disorder    2014-06-29 02:18:00   ...    sunny-side    1    0
4    aggravated-assault    aggravated-assault    2014-06-29 04:17:00   ...
college-view-south-platte    1    0
5 rows × 8 columns
```

By now, you should be comfortable enough with reshaping to answer questions like, *How many crimes happened in a given year?*. But what if we wanted to drill into that analysis and decide how it changed within each OFFENSE_CATEGORY_ID?

Since pd.DataFrame.resample just works with a pd.DatetimeIndex, it cannot be used to help us group by OFFENSE_CATEGORY_ID and REPORTED_DATE. However, the combination of pd.DataFrame.groupby with a pd.Grouper argument can help us express this:

```
df.groupby([
    "OFFENSE_CATEGORY_ID",
    pd.Grouper(key="REPORTED_DATE", freq="YS"),
], observed=True).agg(
    total_crime=pd.NamedAgg(column="IS_CRIME", aggfunc="sum"),
)
```

```
                              total_crime
OFFENSE_CATEGORY_ID    REPORTED_DATE
aggravated-assault     2012-01-01      1707
        2013-01-01     1631
        2014-01-01     1788
        2015-01-01     2007
        2016-01-01     2139
...            ...            ...
```

```
 white-collar-crime      2013-01-01        771
              2014-01-01      1043
              2015-01-01      1319
              2016-01-01      1232
              2017-01-01      1058
90 rows × 1 columns
```

As a technical aside, the observed=True argument suppresses a warning about using categorical data types in a Group By in the pandas 2.x release; future readers may not need to specify this argument, as it will become the default.

To add in the "year over year" component, we can try out the pd.Series.pct_change method, which expresses each record as a percentage of the one directly preceding it:

```
df.groupby([
    "OFFENSE_CATEGORY_ID",
    pd.Grouper(key="REPORTED_DATE", freq="YS"),
], observed=True).agg(
    total_crime=pd.NamedAgg(column="IS_CRIME", aggfunc="sum"),
).assign(
    yoy_change=lambda x: x["total_crime"].pct_change().astype(pd.
Float64Dtype())
).head(10)
```

```
                         total_crime      yoy_change
OFFENSE_CATEGORY_ID      REPORTED_DATE
aggravated-assault       2012-01-01         1707        <NA>
              2013-01-01      1631      -0.044523
              2014-01-01      1788      0.09626
              2015-01-01      2007      0.122483
              2016-01-01      2139      0.06577
              2017-01-01      1689      -0.210379
all-other-crimes         2012-01-01         1999       0.183541
              2013-01-01      9377      3.690845
              2014-01-01      15507     0.653727
              2015-01-01      15729     0.014316
```

Unfortunately, this is not giving us exactly what we want. If you look closely at the first yoy_change value for all-other-crimes, it shows 0.183541. However, this value is taken by dividing 1999 by 1689, with 1689 coming from the aggravated-assault category. By default, pd.Series.pct_change is not doing anything intelligent – it just divides the current row by the former.

Fortunately, there is a way to fix that, by once again using a Group By. Because our OFFENSE_CATEGORY_ ID is the first index level, we can use a second Group By with level=0 and call the .pct_change method on that. This will prevent us from accidentally comparing all-other-crimes to aggravated-assault:

```python
yoy_crime = df.groupby([
    "OFFENSE_CATEGORY_ID",
    pd.Grouper(key="REPORTED_DATE", freq="YS"),
], observed=True).agg(
    total_crime=pd.NamedAgg(column="IS_CRIME", aggfunc="sum"),
).assign(
    yoy_change=lambda x: x.groupby(
        level=0, observed=True
    ).pct_change().astype(pd.Float64Dtype())
)

yoy_crime.head(10)
```

```
                             total_crime    yoy_change
OFFENSE_CATEGORY_ID   REPORTED_DATE
aggravated-assault    2012-01-01       1707    <NA>
                      2013-01-01       1631    -0.044523
                      2014-01-01       1788    0.09626
                      2015-01-01       2007    0.122483
                      2016-01-01       2139    0.06577
                      2017-01-01       1689    -0.210379
all-other-crimes      2012-01-01       1999    <NA>
                      2013-01-01       9377    3.690845
                      2014-01-01      15507    0.653727
                      2015-01-01      15729    0.014316
```

For a more visual representation, we may want to plot out the total crime and year-over-year change side by side for all of our different groups, building off of what we learned about visualizations back in *Chapter 6, Visualization.*

For brevity and to save some visual space, we are just going to plot a few crime types:

```python
crimes = tuple(("aggravated-assault", "arson", "auto-theft"))
fig, axes = plt.subplots(nrows=len(crimes), ncols=2, sharex=True)

for idx, crime in enumerate(crimes):
    crime_df = yoy_crime.loc[crime]
    ax0 = axes[idx][0]
    ax1 = axes[idx][1]
    crime_df.plot(kind="bar", y="total_crime", ax=ax0, legend=False)
    crime_df.plot(kind="bar", y="yoy_change", ax=ax1, legend=False)

    xlabels = [x.year for x in crime_df.index]
    ax0.set_xticklabels(xlabels)
```

```
        ax0.set_title(f"{crime} total")
        ax1.set_xticklabels(xlabels)
        ax1.set_title(f"{crime} YoY")
        ax0.set_xlabel("")
        ax1.set_xlabel("")

    plt.tight_layout()
```

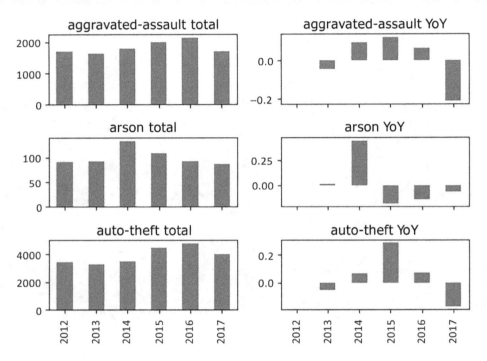

Accurately measuring sensor-collected events with missing values

Missing data can have an immense impact on your data analysis, but it may not always be clear when and to what extent. With detailed and high-volume transactions, it won't always be immediately obvious that a dataset is incomplete. Extra attention must be paid to measure and appropriately impute missing transactions; otherwise, any aggregations performed on such datasets may show an incomplete or even entirely wrong picture of what happened.

For this recipe, we are going to use the *Smart Green Infrastructure Monitoring Sensors - Historical* dataset provided by the Chicago Data Portal. This dataset contains a collection of sensors that measured different environmental factors in the city of Chicago, like water runoff and temperature. In theory, the sensors should have constantly run and reported back values, but in practice, they were prone to intermittent outages that resulted in a loss of data.

How to do it

While the Chicago Data Portal provides the source data as a CSV file spanning the years 2017 and 2018, for this book, we are going to work with a curated Parquet file that only covers the months of June 2017 through October 2017. This alone provides almost 5 million rows of data, which we can load with a simple `pd.read_parquet` call:

```
df = pd.read_parquet(
    "data/sgi_monitoring.parquet",
    dtype_backend="numpy_nullable",
)


df.head()
```

```
     Measurement Title    Measurement Description    Measurement Type    …
Latitude    Longitude    Location
0   UI Labs Bioswale NWS Proba-bility of Precipi-tation  <NA>    TimeWin-
dowBounda-ry    …    41.90715    -87.653996    POINT (-87.653996 41.90715)
1   UI Labs Bioswale NWS Proba-bility of Precipi-tation  <NA>    TimeWin-
dowBounda-ry    …    41.90715    -87.653996    POINT (-87.653996 41.90715)
2   UI Labs Bioswale NWS Proba-bility of Precipi-tation  <NA>    TimeWin-
dowBounda-ry    …    41.90715    -87.653996    POINT (-87.653996 41.90715)
3   UI Labs Bioswale NWS Proba-bility of Precipi-tation  <NA>    TimeWin-
dowBounda-ry    …    41.90715    -87.653996    POINT (-87.653996 41.90715)
4   UI Labs Bioswale NWS Proba-bility of Precipi-tation  <NA>    TimeWin-
dowBounda-ry    …    41.90715    -87.653996    POINT (-87.653996 41.90715)
5 rows × 16 columns
```

The `Measurement Time` column should contain the datetime data for when each event occurred, but upon closer inspection, you will see that pandas did not recognize this as a datetime type:

```
df["Measurement Time"].head()
```

```
0    07/26/2017 07:00:00 AM
1    06/23/2017 07:00:00 AM
2    06/04/2017 07:00:00 AM
3    09/19/2017 07:00:00 AM
4    06/07/2017 07:00:00 AM
Name: Measurement Time, dtype: string
```

As such, the first step in exploring our data will be to convert this to the real datetime type using `pd.to_datetime`. While it isn't clear from the data itself, the Chicago Data Portal documentation notes that these values are local to the Chicago timezone, which we can use `pd.Series.dt.tz_localize` to set:

```
df["Measurement Time"] = pd.to_datetime(
    df["Measurement Time"]
```

```
).dt.tz_localize("US/Central")

df["Measurement Time"]
```

```
0          2017-07-26 07:00:00-05:00
1          2017-06-23 07:00:00-05:00
2          2017-06-04 07:00:00-05:00
3          2017-09-19 07:00:00-05:00
4          2017-06-07 07:00:00-05:00
                     ...
4889976    2017-08-26 20:11:55-05:00
4889977    2017-08-26 20:10:54-05:00
4889978    2017-08-26 20:09:53-05:00
4889979    2017-08-26 20:08:52-05:00
4889980    2017-08-26 20:07:50-05:00
Name: Measurement Time, Length: 4889981, dtype: datetime64[ns, US/Central]
```

As mentioned, this dataset collects feedback from sensors that measure different environmental factors, like water runoff and temperature. Inspecting the Measurement Type and Units column should give us a better idea of what we are looking at:

```
df[["Measurement Type", "Units"]].value_counts()
```

```
Measurement Type        Units
Temperature             degrees Celsius                   721697
DifferentialPressure    pascals                           721671
WindSpeed               meters per second                 721665
Temperature             millivolts                        612313
SoilMoisture            millivolts                        612312
RelativeHumidity        percent                           389424
CumulativePrecipitation count                             389415
WindDirection           degrees from north                389413
SoilMoisture            Percent Volumetric Water Content  208391
CumulativePrecipitation inches                            122762
TimeWindowBoundary      universal coordinated time           918
Name: count, dtype: int64
```

Because the different sensors produce different measurements for different types of data, we must be careful not to compare more than one sensor at a time. For this analysis, we are going to just focus on the TM1 Temp Sensor, which only measures the temperature using a unit of millivolts. Additionally, we are going to single in on one Data Stream ID, which the Chicago Data Portal documents as:

> *An identifier for the measurement type and location. All records with the same value should be comparable.*

```
df[df["Measurement Description"] == "TM1 Temp Sensor"]["Data Stream ID"].value_
counts()
```

```
Data Stream ID
33305    211584
39197    207193
39176    193536
Name: count, dtype: Int64
```

For this analysis, we are going to only look at Data Stream ID 39176. After filtering, we are also going to set Measurement Time as our row index and sort it:

```
mask = (
    (df["Measurement Description"] == "TM1 Temp Sensor")
    & (df["Data Stream ID"] == 39176)
)
df = df[mask].set_index("Measurement Time").sort_index()
df[["Measurement Type", "Units"]].value_counts()
```

```
Measurement Type  Units
Temperature       millivolts    193536
Name: count, dtype: int64
```

The Measurement Value column contains the actual millivolts reading from the sensors. Let's start by resampling to the daily level and using mean aggregation on that column to try and understand our data at a higher level:

```
df.resample("D")["Measurement Value"].mean().plot()
```

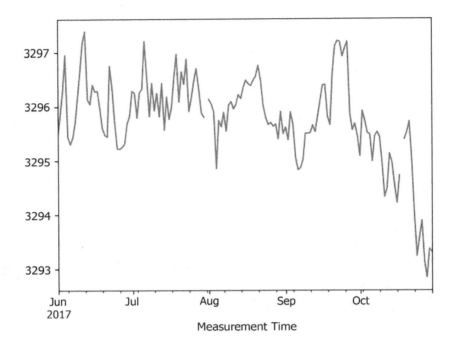

Measurement Time

Almost immediately, we can see some issues with our data. Most notably, there are two gaps where the lines break toward the end of July and middle of October, which are almost assuredly records that were not collected due to the sensors being down.

Let's try narrowing our date range so that we can more clearly see what days are missing from our dataset:

```
df.loc["2017-07-24":"2017-08-01"].resample("D")["Measurement Value"].mean()
```

```
Measurement Time
2017-07-24 00:00:00-05:00       3295.908956
2017-07-25 00:00:00-05:00       3296.152968
2017-07-26 00:00:00-05:00       3296.460156
2017-07-27 00:00:00-05:00       3296.697269
2017-07-28 00:00:00-05:00       3296.328725
2017-07-29 00:00:00-05:00       3295.882705
2017-07-30 00:00:00-05:00       3295.800989
2017-07-31 00:00:00-05:00              <NA>
2017-08-01 00:00:00-05:00       3296.126888
Freq: D, Name: Measurement Value, dtype: Float64
```

As you can see, we have no data collected at all for July 31, 2017. To fix this, we can simply chain in a call to pd.Series.interpolate, which will fill in the missing days with the average of the values directly preceding and following:

```
df.resample("D")["Measurement Value"].mean().interpolate().plot()
```

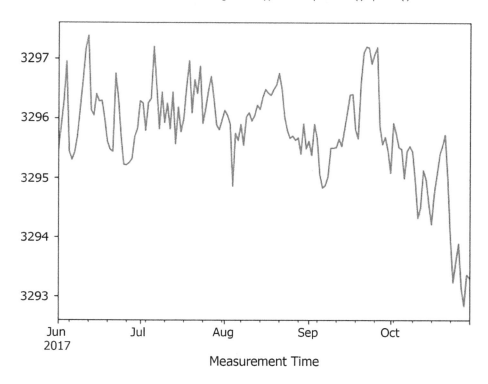

Voila! Now, we no longer have any gaps in our data collection, yielding a visually appealing, fully drawn-in visual.

There's more...

How you handle missing data also depends on the aggregation function that you are using. In this recipe, the mean is a relatively forgiving function; missing transactions can be masked by the fact that they do not materially change the average being produced.

However, if we were looking to measure the daily summation of our readings, we would still have some more work to do. For starters, let's see what a daily-resampled summation of these readings looks like:

```
df.resample("D")["Measurement Value"].sum().plot()
```

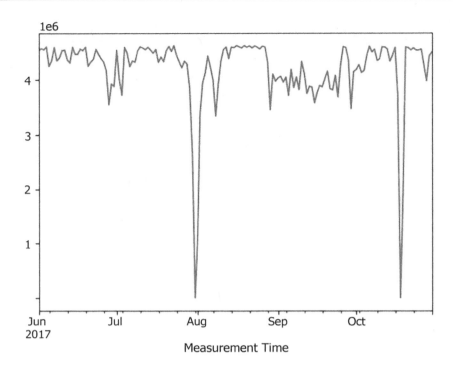

Measurement Time

Things look more dire than in the case of wanting the average. We still see huge dips in late July and October, which we know go back to a lack of data. However, when we dive into the data at the end of July that we saw before, the summation will reveal a few more interesting things about our data:

```
df.loc["2017-07-30 15:45:00":"2017-08-01"].head()
```

```
      Measurement Title   Measurement Description   Measurement Type   …
Latitude   Longitude   Location   Measurement Time
2017-07-30 15:48:44-05:00    Argyle - Thun-der 1: TM1 Temp Sensor    TM1
Temp Sensor    Temperature    …    41.973086    -87.659725    POINT (-87.659725
41.973086)
2017-07-30 15:49:45-05:00    Argyle - Thun-der 1: TM1 Temp Sensor    TM1
Temp Sensor    Temperature    …    41.973086    -87.659725    POINT (-87.659725
41.973086)
2017-07-30 15:50:46-05:00    Argyle - Thun-der 1: TM1 Temp Sensor    TM1
Temp Sensor    Temperature    …    41.973086    -87.659725    POINT (-87.659725
41.973086)
2017-08-01 15:21:33-05:00    Argyle - Thun-der 1: TM1 Temp Sensor    TM1
Temp Sensor    Temperature    …    41.973086    -87.659725    POINT (-87.659725
41.973086)
2017-08-01 15:22:34-05:00    Argyle - Thun-der 1: TM1 Temp Sensor    TM1
Temp Sensor    Temperature    …    41.973086    -87.659725    POINT (-87.659725
41.973086)
5 rows × 15 columns
```

It wasn't just the day of July 31 when we had an outage. The mean aggregation we did before masked the fact that the sensors went down sometime after 15:50:46 on July 30 and did not come back online until 15:21:33 on August 1 – an outage of almost 2 full days.

Another interesting thing to try and measure is the expected frequency with which our data should be populated. From an initial glance at our data, it appears as if each minute should supply a data point, but if you try to measure how many events were collected each hour, you will see a different story:

```
df.resample("h").size().plot()
```

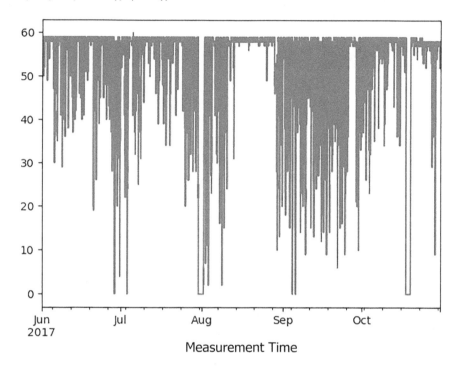

Many of the hourly intervals appear to have close to 60 events collected, although surprisingly, only 1 hour actually collected a full 60:

```
df.resample("h").size().loc[lambda x: x >= 60]
```

```
Measurement Time
2017-07-05 15:00:00-05:00    60
Freq: h, dtype: int64
```

To fix this, let's try once again to resample our data by the minute and interpolate where results are missing:

```
df.resample("min")["Measurement Value"].sum().interpolate()
```

```
Measurement Time
2017-06-01 00:00:00-05:00    3295.0
```

```
2017-06-01 00:01:00-05:00        3295.0
2017-06-01 00:02:00-05:00        3295.0
2017-06-01 00:03:00-05:00        3295.0
2017-06-01 00:04:00-05:00        3295.0
                                  ...
2017-10-30 23:55:00-05:00        3293.0
2017-10-30 23:56:00-05:00        3293.0
2017-10-30 23:57:00-05:00           0.0
2017-10-30 23:58:00-05:00        3293.0
2017-10-30 23:59:00-05:00        3293.0
Freq: min, Name: Measurement Value, Length: 218880, dtype: Float64
```

There is a slight caveat users should be aware of with the summation of missing values. By default, pandas will sum all missing values to 0 instead of a missing value. In the case of our resample to minutes, the data point at 2017-10-30 23:57:00 had no values to sum, so pandas returned the value of 0 instead of a missing value indicator.

We need a missing value indicator for the resample to work. Luckily, we can still get this by providing the sum method with a min_count= argument that is 1 (or greater), essentially establishing how many non-missing values must be seen to yield a non-missing result:

```
interpolated = df.resample("min")["Measurement Value"].sum(min_count=1).
interpolate()
interpolated
```

```
Measurement Time
2017-06-01 00:00:00-05:00        3295.0
2017-06-01 00:01:00-05:00        3295.0
2017-06-01 00:02:00-05:00        3295.0
2017-06-01 00:03:00-05:00        3295.0
2017-06-01 00:04:00-05:00        3295.0
                                  ...
2017-10-30 23:55:00-05:00        3293.0
2017-10-30 23:56:00-05:00        3293.0
2017-10-30 23:57:00-05:00        3293.0
2017-10-30 23:58:00-05:00        3293.0
2017-10-30 23:59:00-05:00        3293.0
Freq: min, Name: Measurement Value, Length: 218880, dtype: Float64
```

As you can see, the value for 2017-10-30 23:57:00 now shows as 3293, which was interpolated by taking both the preceding and following values.

With that out of the way, let's now confirm that we always see 60 events per hour:

```
interpolated.resample("h").size().plot()
```

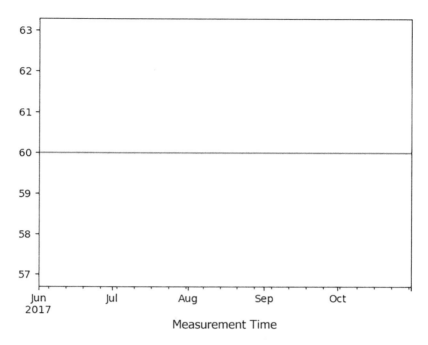

That check looks good, so now, we can try to downsample again back to the daily level and see what the overall summation trend looks like:

```
interpolated.resample("D").sum().plot()
```

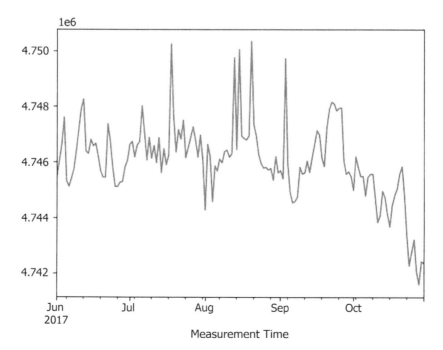

This is a drastically different graph from what we started with. We not only removed the extreme outliers from influencing the *y*-axis of our graph but we can also see a general lift in the lower bounds of values. In our original graph, the lower bound of the total millivolts measured was commonly in the range of 3.5–4 million per day, but now, our lower bound appears somewhere around 4.74 million.

In effect, by paying attention to and handling missing values in our time series data, we were able to yield many different insights from our dataset. In relatively few lines of code, pandas has helped us clearly and concisely get our data to a much better place than where we started.

Join our community on Discord

Join our community's Discord space for discussions with the authors and other readers:

```
https://packt.link/pandas
```

10

General Usage and Performance Tips

At this point in the book, we have covered a rather large part of the pandas library while walking through sample applications to reinforce good usage. Equipped with all of this knowledge, you are now well prepared to step into the real world and start applying everything you have learned to your data analysis problems.

This chapter will offer some tips and tricks you should keep in mind as you go out on your own. The recipes presented in this chapter are common mistakes I see by pandas users of all experience levels. While well-intentioned, improper usage of pandas constructs can leave a lot of performance on the table. When your datasets are smaller in size that may not be a big issue, but data has the tendency to grow, not to retreat in size. Using proper idioms and avoiding the maintenance burden of inefficient code can yield significant time and money savings for your organization.

We are going to cover the following recipes in this chapter:

- Avoid dtype=object
- Be cognizant of data sizes
- Use vectorized functions instead of loops
- Avoid mutating data
- Dictionary-encode low cardinality data
- Test-driven development features

Avoid dtype=object

Using dtype=object to store strings is one of the most error-prone and inefficient things you can do in pandas. Unfortunately, for the longest time, dtype=object was the only way to work with string data; this wasn't "solved" until the 1.0 release.

I intentionally put "solved" in quotes because, while pandas 1.0 did introduce the pd.StringDtype(), it was not used by default by many construction and I/O methods until the 3.0 release. In effect, unless you told pandas otherwise, you would end up with dtype=object for all your string data in the 2.x series. For what it's worth, the pd.StringDtype() that was introduced in 1.0 helped to assert you *only* stored strings, but it was never optimized for performance until the pandas 3.0 release.

If you are using the 3.0 release of pandas and beyond, chances are you will still come across legacy code that reads like ser = ser.astype(object). More often than not, such calls should be replaced with ser = ser.astype(pd.StringDtype()), unless you truly do need to store Python objects in a pd.Series. Unfortunately, there is no true way to know the intent, so you as a developer should be aware of the pitfalls of using dtype=object and how to identify if it can suitably be replaced with the pd.StringDtype().

How to do it

We already covered some of the issues with using dtype=object back in *Chapter 3, Data Types,* but it is worth restating and expanding upon some of those issues here.

For an easy comparison, let's create two pd.Series objects with identical data, where one uses the object data type and the other uses the pd.StringDtype:

```
ser_obj = pd.Series(["foo", "bar", "baz"] * 10_000, dtype=object)
ser_str = pd.Series(["foo", "bar", "baz"] * 10_000, dtype=pd.StringDtype())
```

> 💡 **Quick tip:** Enhance your coding experience with the **AI Code Explainer** and **Quick Copy** features. Open this book in the next-gen Packt Reader. Click the **Copy** button (1) to quickly copy code into your coding environment, or click the **Explain** button (2) to get the AI assistant to explain a block of code to you.
>
>
> ```
>                                                              Copy      Explain
> function calculate(a, b) {
>   return {sum: a + b};                                        1          2
> };
> ```
>
> 🔒 **The next-gen Packt Reader** is included for free with the purchase of this book. Unlock it by scanning the QR code below or visiting https://www.packtpub.com/unlock/9781836205876.
>
>

Attempting to assign a non-string value to `ser_str` will fail:

```
ser_str.iloc[0] = False
```

```
TypeError: Cannot set non-string value 'False' into a StringArray.
```

By contrast, the object-typed `pd.Series` will gladly accept our `Boolean` value:

```
ser_obj.iloc[0] = False
```

In turn, this just ends up obfuscating where issues with your data may occur. With `pd.StringDtype`, the point of failure was very obvious when we tried to assign non-string data. With the object data type, you may not discover there is a problem until later in your code, when you try some string operation like capitalization:

```
ser_obj.str.capitalize().head()
```

```
0     NaN
1     Bar
2     Baz
3     Foo
4     Bar
dtype: object
```

Instead of raising an error, pandas just decided to set our `False` entry in the first row to a missing value. Odds are just silently setting things to missing values like that is not the behavior you wanted, but with the object data type, you lose a lot of control over your data quality.

If you are working with pandas 3.0 and beyond, you will also see that, when PyArrow is installed, `pd.StringDtype` becomes significantly faster. Let's recreate our `pd.Series` objects to measure this:

```
ser_obj = pd.Series(["foo", "bar", "baz"] * 10_000, dtype=object)
ser_str = pd.Series(["foo", "bar", "baz"] * 10_000, dtype=pd.StringDtype())
```

For a quick timing comparison, let's use the `timeit` module, built into the standard library:

```
import timeit
timeit.timeit(ser_obj.str.upper, number=1000)
```

```
2.2286621460007154
```

Compare that runtime to the same values but with the proper `pd.StringDtype`:

```
timeit.timeit(ser_str.str.upper, number=1000)
```

```
2.7227514309997787
```

Unfortunately, users prior to the 3.0 release will not see any performance difference, but the data validation alone is worth it to move away from `dtype=object`.

So what is the easiest way to avoid `dtype=object`? If you are fortunate enough to be working with the 3.0 release and beyond of pandas, you will naturally not run into this data type as often, as a natural evolution in the library. Even still, and for users that may still be using the pandas 2.x series, I advise using the `dtype_backend="numpy_nullable"` argument with I/O methods:

```python
import io
data = io.StringIO("int_col,string_col\n0,foo\n1,bar\n2,baz")

data.seek(0)
pd.read_csv(data, dtype_backend="numpy_nullable").dtypes
```

```
int_col                  Int64
string_col       string[python]
dtype: object
```

If you are constructing a `pd.DataFrame` by hand, you can use `pd.DataFrame.convert_dtypes` paired with the same `dtype_backend="numpy_nullable"` argument:

```python
df = pd.DataFrame([
    [0, "foo"],
    [1, "bar"],
    [2, "baz"],
], columns=["int_col", "string_col"])
df.convert_dtypes(dtype_backend="numpy_nullable").dtypes
```

```
int_col                  Int64
string_col       string[python]
dtype: object
```

Please note that the `numpy_nullable` term is a bit of a misnomer. The argument would have probably been better named `pandas_nullable` or even just `pandas` or `nullable`, but when it was first introduced, it was strongly tied to the NumPy system still. Over time, the term `numpy_nullable` stuck, but the types moved away from using NumPy. Beyond the publication of this book, there may be a more suitable value to use to get the same behavior, which essentially asks for optimal data types in pandas that can support missing values.

While using `dtype=object` is most commonly *misused* for strings, it also exposes some rough edges with datetimes. I commonly see code like this from new users trying to create what they think is a `pd.Series` of dates:

```python
import datetime

ser = pd.Series([
    datetime.date(2024, 1, 1),
    datetime.date(2024, 1, 2),
```

```
        datetime.date(2024, 1, 3),
    ])

ser
```

```
0      2024-01-01
1      2024-01-02
2      2024-01-03
dtype: object
```

While this is a logical way to try and accomplish the task at hand, the problem is that pandas does not have a true *date* type. Instead, these get stored in an object data type array using the datetime.date type from the Python standard library. This rather unfortunate usage of Python objects obfuscates the fact that you are trying to work with dates, and subsequently trying to use the pd.Series.dt accessor will throw an error:

```
ser.dt.year
```

```
AttributeError: Can only use .dt accessor with datetimelike values
```

Back in *Chapter 3*, *Data Types*, we talked briefly about the PyArrow date32 type, which would be a more native solution to this problem:

```
import datetime

ser = pd.Series([
    datetime.date(2024, 1, 1),
    datetime.date(2024, 1, 2),
    datetime.date(2024, 1, 3),
], dtype=pd.ArrowDtype(pa.date32()))

ser
```

```
0      2024-01-01
1      2024-01-02
2      2024-01-03
dtype: date32[day][pyarrow]
```

This will then unlock the pd.Series.dt attributes for use:

```
ser.dt.year
```

```
0      2024
1      2024
2      2024
dtype: int64[pyarrow]
```

I find this nuance rather unfortunate and hope future versions of pandas will be able to abstract these issues away, but nonetheless, they are present at the time of publication and may be for some time.

In spite of all of the downsides that I have highlighted with respect to dtype=object, it still does have its uses when dealing with messy data. Sometimes, you may not know anything about your data and just need to inspect it before making further decisions. The object data type gives you the flexibility to load essentially any data and apply the same pandas algorithms to it. While these algorithms may not be very efficient, they still give you a consistent way to interact with and explore your data, ultimately buying you time to figure out how to best cleanse it and store it in a more proper data form. For this reason, I consider dtype=object best as a staging area – I would not advise keeping your types in it, but the fact that it buys you time to make assertions about your data types can be an asset.

Be cognizant of data sizes

As your datasets grow larger, you may find that you have to pick more optimal data types to ensure your pd.DataFrame can still fit into memory.

Back in *Chapter 3, Data Types*, we discussed the different integral types and how they are a trade-off between memory usage and capacity. When dealing with untyped data sources like CSV and Excel files, pandas will err on the side of using too much memory as opposed to picking the wrong capacity. This conservative approach can lead to inefficient usage of your system's memory, so knowing how to optimize that can make the difference between loading a file and receiving an OutOfMemory error.

How to do it

To illustrate the impact of picking proper data types, let's start with a relatively large pd.DataFrame composed of Python integers:

```python
df = pd.DataFrame({
    "a": [0] * 100_000,
    "b": [2 ** 8] * 100_000,
    "c": [2 ** 16] * 100_000,
    "d": [2 ** 32] * 100_000,
})
df = df.convert_dtypes(dtype_backend="numpy_nullable")

df.head()
```

```
   a    b      c           d
0  0  256  65536  4294967296
1  0  256  65536  4294967296
2  0  256  65536  4294967296
3  0  256  65536  4294967296
4  0  256  65536  4294967296
```

With the integral types, determining how much memory each pd.Series requires is a rather simple exercise. With a pd.Int64Dtype, each record is a 64-bit integer that requires 8 bytes of memory. Alongside each record, the pd.Series associates a single byte that is either 0 or 1, telling us if the record is missing or not. Thus, in total, we need 9 bytes for each record, and with 100,000 records per pd.Series, our memory usage should come out to 900,000 bytes. pd.DataFrame.memory_usage confirms that this math is correct:

```
df.memory_usage()
```

```
Index         128
a          900000
b          900000
c          900000
d          900000
dtype: int64
```

If you know what the types should be, you could explicitly pick better sizes for the pd.DataFrame columns using .astype:

```
df.assign(
    a=lambda x: x["a"].astype(pd.Int8Dtype()),
    b=lambda x: x["b"].astype(pd.Int16Dtype()),
    c=lambda x: x["c"].astype(pd.Int32Dtype()),
).memory_usage()
```

```
Index         128
a          200000
b          300000
c          500000
d          900000
dtype: int64
```

As a convenience, pandas can try and infer better sizes for you with a call pd.to_numeric. Passing the downcast="signed" argument will ensure that we continue to work with signed integers, and we will continue to pass dtype_backend="numpy_nullable" to ensure we get proper missing value support:

```
df.select_dtypes("number").assign(
    **{x: pd.to_numeric(
        y, downcast="signed", dtype_backend="numpy_nullable"
    ) for x, y in df.items()}
).memory_usage()
```

```
Index         128
a          200000
b          300000
```

```
c           500000
d           900000
dtype: int64
```

Use vectorized functions instead of loops

Python as a language is celebrated for its looping prowess. Whether you are working with a list or a dictionary, looping over an object in Python is a relatively easy task to perform, and can allow you to write really clean, concise code.

Even though pandas is a Python library, those same looping constructs are ironically an impediment to writing idiomatic, performant code. In contrast to looping, pandas offers *vectorized computations*, i.e, computations that work with all of the elements contained within a pd.Series but which do not require you to explicitly loop.

How to do it

Let's start with a simple pd.Series constructed from a range:

```
ser = pd.Series(range(100_000), dtype=pd.Int64Dtype())
```

We could use the built-in pd.Series.sum method to easily calculate the summation:

```
ser.sum()
```

```
4999950000
```

Looping over the pd.Series and accumulating your own result will yield the same number:

```
result = 0
for x in ser:
    result += x

result
```

```
4999950000
```

Yet the two code samples are nothing alike. With pd.Series.sum, pandas performs the summation of elements in a lower-level language like C, avoiding any interaction with the Python runtime. In pandas speak, we would refer to this as a *vectorized* function.

By contrast, the for loop is handled by the Python runtime, and as you may or may not be aware, Python is a much slower language than C.

To put some tangible numbers forth, we can run a simple timing benchmark using Python's timeit module. Let's start with pd.Series.sum:

```
timeit.timeit(ser.sum, number=1000)
```

```
0.04479526499926578
```

Let's compare that to the Python loop:

```
def loop_sum():
    result = 0
    for x in ser:
        result += x


timeit.timeit(loop_sum, number=1000)
```

```
5.392715779991704
```

That's a huge slowdown with the loop!

Generally, you should look to use the built-in vectorized functions of pandas for most of your analysis needs. For more complex applications, reach for the `.agg`, `.transform`, `.map`, and `.apply` methods, which were covered back in *Chapter 5, Algorithms and How to Apply Them*. You should be able to avoid using `for` loops in 99.99% of your analyses; if you find yourself using them more often, you should rethink your design, more than likely after a thorough re-read of *Chapter 5, Algorithms and How to Apply Them*.

The one exception to this rule where it may make sense to use a `for` loop is when dealing with a `pd.GroupBy` object, which can be efficiently iterated like a dictionary:

```
df = pd.DataFrame({
    "column": ["a", "a", "b", "a", "b"],
    "value": [0, 1, 2, 4, 8],
})
df = df.convert_dtypes(dtype_backend="numpy_nullable")


for label, group in df.groupby("column"):
    print(f"The group for label {label} is:\n{group}\n")
```

```
The group for label a is:
  column  value
0      a      0
1      a      1
3      a      4

The group for label b is:
  column  value
2      b      2
4      b      8
```

Avoid mutating data

Although pandas allows you to mutate data, the cost impact of doing so varies by data type. In some cases, it can be prohibitively expensive, so you will be best served trying to minimize mutations you have to perform at all costs.

How to do it

When thinking about data mutation, a best effort should be made to mutate before loading into a pandas structure. We can easily illustrate a performance difference by comparing the time to mutate a record after loading it into a pd.Series:

```python
def mutate_after():
    data = ["foo", "bar", "baz"]
    ser = pd.Series(data, dtype=pd.StringDtype())
    ser.iloc[1] = "BAR"


timeit.timeit(mutate_after, number=1000)
```

```
0.041951814011554234
```

To the time it takes if the mutation was performed beforehand:

```python
def mutate_before():
    data = ["foo", "bar", "baz"]
    data[1] = "BAR"
    ser = pd.Series(data, dtype=pd.StringDtype())


timeit.timeit(mutate_before, number=1000)
```

```
0.019495725005981512
```

There's more...

You can go down a technical rabbit hole trying to decipher the impact of mutating various data types in pandas, across all of the different versions. However, starting in pandas 3.0, the behavior started to become more consistent with the introduction of Copy-on-Write, which was proposed as part of PDEP-07. In essence, any time you try to mutate a pd.Series or pd.DataFrame, you end up with a copy of the original data.

While this behavior is now easier to anticipate, it also means that mutations are potentially very expensive, especially if you try to mutate a large pd.Series or pd.DataFrame.

Dictionary-encode low cardinality data

Back in *Chapter 3, Data Types,* we talked about the categorical data type, which can help to reduce memory usage by replacing occurrences of strings (or any data type really) with much smaller integral code. While *Chapter 3, Data Types,* provides a good technical deep dive, it is worth restating this as a best practice here, given how significant of a saving this can represent when working with *low cardinality* data, i.e, data where the ratio of unique values to the overall record count is relatively low.

How to do it

Just to drive home the point about memory savings, let's create a *low cardinality* pd.Series. Our pd.Series is going to have 300,000 rows of data, but only three unique values of "foo", "bar", and "baz":

```
values = ["foo", "bar", "baz"]
ser = pd.Series(values * 100_000, dtype=pd.StringDtype())
ser.memory_usage()
```

```
2400128
```

Simply changing this to a categorical data type will yield massive memory improvements:

```
cat = pd.CategoricalDtype(values)
ser = pd.Series(values * 100_000, dtype=cat)
ser.memory_usage()
```

```
300260
```

Test-driven development features

Test-driven development (or **TDD**, for short) is a popular software development practice that aims to improve code quality and maintenance. At a high level, TDD starts with a developer creating tests that describe the expected functionality of their change. The tests start in a failed state, and the developer can become confident that their implementation is correct when the tests finally pass.

Tests are often the first thing code reviewers look at when considering code changes (when contributing to pandas, tests are a must!). After a change with an accompanying test has been accepted, that same test will be re-run for any subsequent code changes, ensuring that your code base continues to work as expected over time. Generally, properly constructed tests can help your code base scale out, while mitigating the risk of regressions as you develop new features.

The pandas library exposes utilities that make writing tests for your pd.Series and pd.DataFrame objects possible through the pd.testing module, which we will review in this recipe.

How it works

The Python standard library offers the unittest module to declare and automate the execution of your tests. To create tests, you typically create a class that inherits from unittest.TestCase, and create methods on that class that make assertions about your program behavior.

In the following code sample, the `MyTests.test_42` method is going to call `unittest.TestCase.assertEqual` with two arguments, `21 * 2` and `42`. Since those arguments are logically equal, the test execution will pass:

```python
import unittest

class MyTests(unittest.TestCase):

    def test_42(self):
        self.assertEqual(21 * 2, 42)

def suite():
    suite = unittest.TestSuite()
    suite.addTest(MyTests("test_42"))
    return suite

runner = unittest.TextTestRunner()
runner.run(suite())
```

```
.
----------------------------------------------------------------------
Ran 1 test in 0.001s

OK
<unittest.runner.TextTestResult run=1 errors=0 failures=0>
```

Now let's try to follow that same execution framework with pandas, but instead of comparing `21 * 2` to `42`, we are going to try and compare two `pd.Series` objects:

```python
def some_cool_numbers():
    return pd.Series([42, 555, pd.NA], dtype=pd.Int64Dtype())

class MyTests(unittest.TestCase):

    def test_cool_numbers(self):
        result = some_cool_numbers()
        expected = pd.Series([42, 555, pd.NA], dtype=pd.Int64Dtype())
        self.assertEqual(result, expected)

def suite():
    suite = unittest.TestSuite()
    suite.addTest(MyTests("test_cool_numbers"))
```

```
    return suite

runner = unittest.TextTestRunner()
runner.run(suite())
```

```
E
======================================================================
ERROR: test_cool_numbers (__main__.MyTests)
----------------------------------------------------------------------
Traceback (most recent call last):
 File "/tmp/ipykernel_79586/2361126771.py", line 9, in test_cool_numbers
   self.assertEqual(result, expected)
 File "/usr/lib/python3.9/unittest/case.py", line 837, in assertEqual
   assertion_func(first, second, msg=msg)
 File "/usr/lib/python3.9/unittest/case.py", line 827, in _baseAssertEqual
   if not first == second:
 File "/home/willayd/clones/Pandas-Cookbook-Third-Edition/lib/python3.9/site-
packages/pandas/core/generic.py", line 1577, in __nonzero__
   raise ValueError(
ValueError: The truth value of a Series is ambiguous. Use a.empty, a.bool(),
a.item(), a.any() or a.all().
----------------------------------------------------------------------
Ran 1 test in 0.004s

FAILED (errors=1)
<unittest.runner.TextTestResult run=1 errors=1 failures=0>
```

Well...that was surprising!

The underlying issue here is that the call to `self.assertEqual(result, expected)` executes the expression `result == expected`. If the result of that expression were `True`, the test would pass; an expression that returns `False` would fail the test.

However, pandas overloads the equality operator for a `pd.Series`, so that instead of returning `True` or `False`, you actually get back another `pd.Series` with an element-wise comparison:

```
result = some_cool_numbers()
expected = pd.Series([42, 555, pd.NA], dtype=pd.Int64Dtype())
result == expected
```

```
0    True
1    True
2    <NA>
dtype: boolean
```

Since testing frameworks don't know what to make of this, you will have to reach for custom functions in the pd.testing namespace. For pd.Series comparison, pd.testing.assert_series_equal is the right tool for the job:

```python
import pandas.testing as tm

def some_cool_numbers():
    return pd.Series([42, 555, pd.NA], dtype=pd.Int64Dtype())

class MyTests(unittest.TestCase):

    def test_cool_numbers(self):
        result = some_cool_numbers()
        expected = pd.Series([42, 555, pd.NA], dtype=pd.Int64Dtype())
        tm.assert_series_equal(result, expected)

def suite():
    suite = unittest.TestSuite()
    suite.addTest(MyTests("test_cool_numbers"))
    return suite

runner = unittest.TextTestRunner()
runner.run(suite())
```

```
.
----------------------------------------------------------------------
Ran 1 test in 0.001s

OK
<unittest.runner.TextTestResult run=1 errors=0 failures=0>
```

For completeness, let's trigger an intentional failure and review the output:

```python
def some_cool_numbers():
    return pd.Series([42, 555, pd.NA], dtype=pd.Int64Dtype())

class MyTests(unittest.TestCase):

    def test_cool_numbers(self):
        result = some_cool_numbers()
        expected = pd.Series([42, 555, pd.NA], dtype=pd.Int32Dtype())
        tm.assert_series_equal(result, expected)
```

```python
def suite():
    suite = unittest.TestSuite()
    suite.addTest(MyTests("test_cool_numbers"))
    return suite


runner = unittest.TextTestRunner()
runner.run(suite())
```

```
F
================================================================
FAIL: test_cool_numbers (__main__.MyTests)
----------------------------------------------------------------
Traceback (most recent call last):
  File "/tmp/ipykernel_79586/2197259517.py", line 9, in test_cool_numbers
    tm.assert_series_equal(result, expected)
  File "/home/willayd/clones/Pandas-Cookbook-Third-Edition/lib/python3.9/site-
packages/pandas/_testing/asserters.py", line 975, in assert_series_equal
    assert_attr_equal("dtype", left, right, obj=f"Attributes of {obj}")
  File "/home/willayd/clones/Pandas-Cookbook-Third-Edition/lib/python3.9/site-
packages/pandas/_testing/asserters.py", line 421, in assert_attr_equal
    raise_assert_detail(obj, msg, left_attr, right_attr)
  File "/home/willayd/clones/Pandas-Cookbook-Third-Edition/lib/python3.9/site-
packages/pandas/_testing/asserters.py", line 614, in raise_assert_detail
    raise AssertionError(msg)
AssertionError: Attributes of Series are different

Attribute "dtype" are different
[left]:  Int64
[right]: Int32
----------------------------------------------------------------
Ran 1 test in 0.003s

FAILED (failures=1)
<unittest.runner.TextTestResult run=1 errors=0 failures=1>
```

Within the test failure traceback, pandas is telling us that the data types of the compared objects are not the same. The result of a call to some_cool_numbers returns a pd.Series with a pd.Int64Dtype, whereas our expectation was looking for a pd.Int32Dtype.

While these examples focused on using pd.testing.assert_series_equal, the equivalent method for a pd.DataFrame is pd.testing.assert_frame_equal. Both of these functions know how to handle potentially different row indexes, column indexes, values, and missing value semantics, and will report back informative errors to the test runner if expectations are not met.

There's more...

This recipe used the unittest module because it is built into the Python language. However, many large Python projects, particularly in the scientific Python space, use the pytest library to write and execute unit tests.

In contrast to unittest, pytest abandons a class-based testing structure with setUp and tearDown methods, opting instead for a test fixture-based approach. A comparison of these two different testing paradigms can be found within the *pytest* documentation.

The pytest library also offers a rich set of plugins. Some plugins may aim to improve integration with third-party libraries (as is the case for pytest-django and pytest-sqlalchemy), whereas others may be focused on scaling your test suite to use all of your system's resources (as is the case for pytest-xdist). There are countless plugin use cases in between, so I strongly recommend giving pytest and its plugin ecosystem a look for testing your Python code bases.

Unlock this book's exclusive benefits now

This book comes with additional benefits designed to elevate your learning experience.

Note: Have your purchase invoice ready before you begin. https://www.packtpub.com/unlock/9781836205876

11

The pandas Ecosystem

While the pandas library offers an impressive array of features, its popularity owes much to the vast amount of third-party libraries that work with it in a complementary fashion. We cannot hope to cover all of those libraries in this chapter, nor can we even dive too deep into how any individual library works. However, just knowing these tools exist and understanding what they offer can serve as a great inspiration for future learning.

While pandas is an amazing tool, it has its flaws, which we have tried to highlight throughout this book; pandas cannot hope to solve every analytical problem there is. I strongly encourage you to get familiar with the tools outlined in this chapter and to also refer to the pandas ecosystem documentation (`https://pandas.pydata.org/about/`) when looking for new and specialized tools.

As a technical note on this chapter, it is possible that these code blocks may break or change behavior as new releases of the libraries are released. While we went to great lengths throughout this book to try and write pandas code that is "future-proof", it becomes more difficult to guarantee that as we write about third party dependencies (and their dependencies). If you encounter any issues running the code in this chapter, be sure to reference the `requirements.txt` file provided alongside the code samples with this book. That file will contain a list of dependencies and versions that are known to work with this chapter.

We will cover the following recipes in this chapter:

- Foundational libraries
- Exploratory data analysis
- Data validation
- Visualization
- Data science
- Databases
- Other DataFrame libraries

Foundational libraries

Like many open source libraries, pandas builds functionality on top of other foundational libraries, letting them manage lower-level details while pandas offers more user-friendly functionality. If you find yourself wanting to dive deeper into technical details beyond what you learn with pandas, these are the libraries you'll want to focus on.

NumPy

NumPy labels itself as the *fundamental package for scientific computing with Python*, and it is the library on top of which pandas was originally built. NumPy is actually an *n*-dimensional library, so you are not limited to two-dimensional data like we get with a pd.DataFrame (pandas actually used to offer 3-d and 4-d panel structures, but they are now long gone).

Throughout this book, we have shown you how to construct pandas objects from NumPy objects, as you can see in the following pd.DataFrame constructor:

```python
arr = np.arange(1, 10).reshape(3, -1)
df = pd.DataFrame(arr)

df
```

```
     0    1    2
0    1    2    3
1    4    5    6
2    7    8    9
```

However, you can also create NumPy arrays from pd.DataFrame objects by using the pd.DataFrame.to_numpy method:

```python
df.to_numpy()
```

```
array([[1, 2, 3],
       [4, 5, 6],
       [7, 8, 9]])
```

Many NumPy functions accept a pd.DataFrame as an argument and will still even return a pd.DataFrame:

```python
np.log(df)
```

```
     0           1           2
0    0.000000    0.693147    1.098612
1    1.386294    1.609438    1.791759
2    1.945910    2.079442    2.197225
```

The main thing to keep in mind with NumPy is that its interoperability with pandas will degrade the moment you need missing values in non-floating point types, or more generally when you try to use data types that are neither integral nor floating point.

The exact rules for this are too complicated to list in this book, but generally, I would advise against ever calling pd.Series.to_numpy or pd.DataFrame.to_numpy for anything other than floating point and integral data.

PyArrow

The other main library that pandas is built on top of is Apache Arrow, which labels itself as a *cross-language development platform for in-memory analytics*. Started by Wes McKinney (the creator of pandas) and announced in his influential Apache Arrow and the *10 Things I Hate About pandas* post (https://wesmckinney.com/blog/apache-arrow-pandas-internals/), the Apache Arrow project defines the memory layout for one-dimensional data structures in a way that allows different languages, programs, and libraries to work with the same data. In addition to defining these structures, the Apache Arrow project offers a vast suite of tooling for libraries to implement the Apache Arrow specifications.

An implementation of Apache Arrow in Python, PyArrow, has been used in particular instances throughout this book. While pandas does not expose a method to convert a pd.DataFrame into PyArrow, the PyArrow library offers a pa.Table.from_pandas method for that exact purpose:

```
tbl = pa.Table.from_pandas(df)
tbl
```

```
pyarrow.Table
0: int64
1: int64
2: int64
----
0: [[1,4,7]]
1: [[2,5,8]]
2: [[3,6,9]]
```

PyArrow similarly offers a pa.Table.to_pandas method to get you from a pa.Table into a pd.DataFrame:

```
tbl.to_pandas()
```

```
    0   1   2
0   1   2   3
1   4   5   6
2   7   8   9
```

Generally, PyArrow is considered a lower-level library than pandas. It mostly aims to serve other library authors more than it does general users looking for a DataFrame library, so, unless you are authoring a library, you may not often need to convert to PyArrow from a pd.DataFrame. However, as the Apache Arrow ecosystem grows, the fact that pandas and PyArrow can interoperate opens up a world of integration opportunities for pandas with many other analytical libraries and databases.

Exploratory data analysis

Oftentimes, you will find yourself provided with a dataset that you know very little about. Throughout this book, we've shown ways to manually sift through data, but there are also tools out there that can help automate potentially tedious tasks and help you grasp the data in a shorter amount of time.

YData Profiling

YData Profiling bills itself as the *"leading package for data profiling, that automates and standardizes the generation of detailed reports, complete with statistics and visualizations."* While we discovered how to manually explore data back in the chapter on visualization, this package can be used as a quick-start to automatically generate many useful reports and features.

To compare this to some of the work we did in those chapters, let's take another look at the vehicles dataset. For now, we are just going to pick a small subset of columns to keep our YData Profiling minimal; for large datasets, the performance can often degrade:

```python
df = pd.read_csv(
    "data/vehicles.csv.zip",
    dtype_backend="numpy_nullable",
    usecols=[
        "id",
        "engId",
        "make",
        "model",
        "cylinders",
        "city08",
        "highway08",
        "year",
        "trany",
    ]
)
df.head()
```

```
     city08   cylinders   engId   ...   model               trany           year
0    19       4           9011    ...   Spider Veloce 2000  Manual 5-spd    1985
1    9        12          22020   ...   Testarossa          Manual 5-spd    1985
2    23       4           2100    ...   Charger             Manual 5-spd    1985
3    10       8           2850    ...   B150/B250 Wagon 2WD Automatic 3-spd 1985
4    17       4           66031   ...   Legacy AWD Turbo    Manual 5-spd    1993
5 rows × 9 columns
```

YData Profiling allows you to easily create a profile report, which contains many common visualizations and helps describe the columns you are working with in your pd.DataFrame.

This book was written using ydata_profiling version 4.9.0. To create the profile report, simply run:

```
from ydata_profiling import ProfileReport
profile = ProfileReport(df, title="Vehicles Profile Report")
```

If running code within a Jupyter notebook, you can see the output of this directly within the notebook with a call to:

```
profile.to_widgets()
```

If you are not using Jupyter, you can alternatively export that profile to a local HTML file and open it from there:

```
profile.to_file("vehicles_profile.html")
```

When looking at the profile, the first thing you will see is a high-level **Overview** section that lists the number of cells with missing data, number of duplicate rows, etc.:

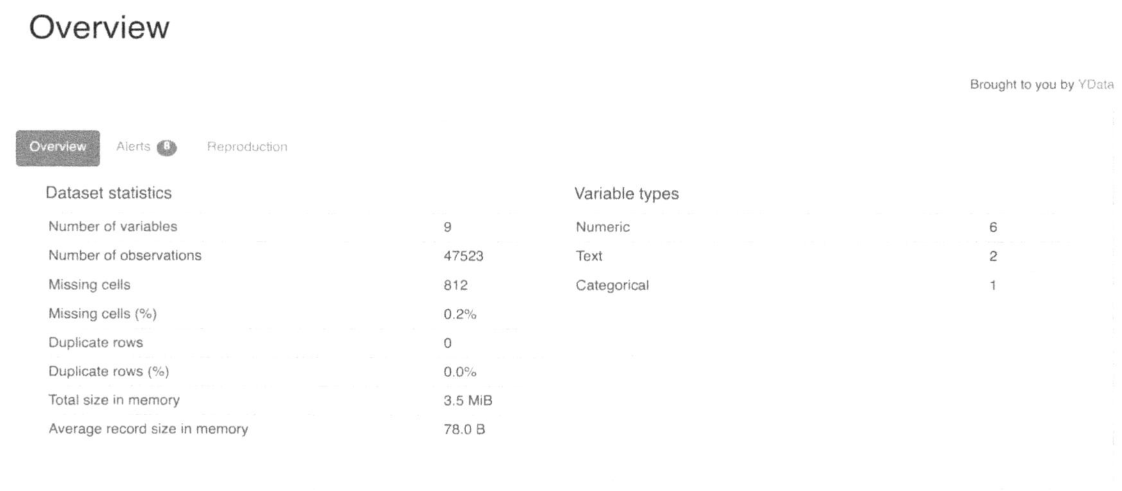

Overview

Brought to you by YData

Overview	Alerts	Reproduction

Dataset statistics		Variable types	
Number of variables	9	Numeric	6
Number of observations	47523	Text	2
Missing cells	812	Categorical	1
Missing cells (%)	0.2%		
Duplicate rows	0		
Duplicate rows (%)	0.0%		
Total size in memory	3.5 MiB		
Average record size in memory	78.0 B		

Figure 11.1: Overview provided by YData Profiling

Each column from your `pd.DataFrame` will be detailed. In the case of a column with continuous values, YData Profiling will create a histogram for you:

Variables

Figure 11.2: Histogram generated by YData Profiling

For categorical variables, the tool will generate a word cloud visualization:

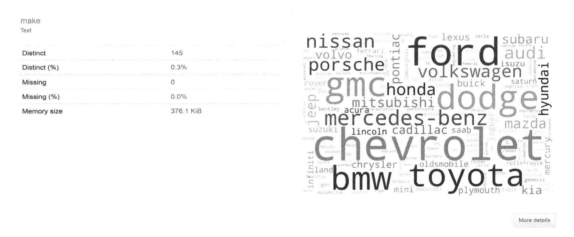

Figure 11.3: Word cloud generated by YData Profiling

To understand how your continuous variables may or may not be correlated, the profile contains a very concise heat map that colors each pair accordingly:

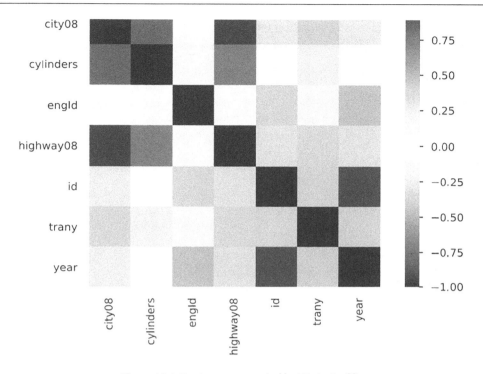

Figure 11.4: Heat map generated by YData Profiling

 Quick tip: Need to see a high-resolution version of this image? Open this book in the next-gen Packt Reader or view it in the PDF/ePub copy.

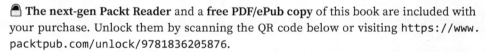 The next-gen Packt Reader and a free PDF/ePub copy of this book are included with your purchase. Unlock them by scanning the QR code below or visiting https://www.packtpub.com/unlock/9781836205876.

While you still will likely need to dive further into your datasets than what this library provides, it can be a great starting point and can help automate the generation of otherwise tedious plots.

Data validation

The "garbage in, garbage out" principle in computing says that no matter how great your code may be, if you start with poor-quality data, your analysis will yield poor-quality results. All too often, data practitioners struggle with issues like unexpected missing data, duplicate values, and broken relationships between modeling entities.

Fortunately, there are tools to help you automate both the data that is input to and output from your models, which ensures trust in the work that you are performing. In this recipe, we are going to look at Great Expectations.

Great Expectations

This book was written using Great Expectations version 1.0.2. To get started, let's once again look at our vehicles dataset:

```
df = pd.read_csv(
    "data/vehicles.csv.zip",
    dtype_backend="numpy_nullable",
    dtype={
        "rangeA": pd.StringDtype(),
        "mfrCode": pd.StringDtype(),
        "c240Dscr": pd.StringDtype(),
        "c240bDscr": pd.StringDtype()
    }
)
df.head()
```

```
    barrels08  barrelsA08  charge120  ...  phevCity  phevHwy  phevComb
0   14.167143  0.0         0.0        ...  0         0        0
1   27.046364  0.0         0.0        ...  0         0        0
2   11.018889  0.0         0.0        ...  0         0        0
3   27.046364  0.0         0.0        ...  0         0        0
4   15.658421  0.0         0.0        ...  0         0        0
5 rows × 84 columns
```

There are a few different ways to use Great Expectations, not all of which can be documented in this cookbook. For the sake of having a self-contained example, we are going to set up and process all of our expectations in memory.

To do this, we are going to import the great_expectations library and create a context for our tests:

```
import great_expectations as gx
context = gx.get_context()
```

Within the context, you can create a data source and a data asset. For non-DataFrame sources like SQL, the data source would typically contain connection credentials, but with the pd.DataFrame residing in memory there is less work to do. The data asset is a grouping mechanism for results. Here we are just creating one data asset, but in real-life use cases you may decide that you want multiple assets to store and organize the validation results that Great Expectations outputs:

```
datasource = context.data_sources.add_pandas(name="pandas_datasource")
data_asset = datasource.add_dataframe_asset(name="vehicles")
```

From there, you can create a batch definition within Great Expectations. For non-DataFrame sources, the batch definition would tell the library how to retrieve data from the source. In the case of pandas, the batch definition will simply retrieve all of the data from the associated pd.DataFrame:

```
batch_definition_name = "dataframe_definition"
batch_definition = data_asset.add_batch_definition_whole_dataframe(
    batch_definition_name
)
batch = batch_definition.get_batch(batch_parameters={
    "dataframe": df
})
```

At this point, you can start to make assertions about the data. For instance, you can use Great Expectations to ensure that a column does not contain any null values:

```
city_exp = gx.expectations.ExpectColumnValuesToNotBeNull(
    column="city08"
)
result = batch.validate(city_exp)
print(result)
```

```
{
  "success": true,
  "expectation_config": {
    "type": "expect_column_values_to_not_be_null",
    "kwargs": {
      "batch_id": "pandas_datasource-vehicles",
      "column": "city08"
    },
    "meta": {}
  },
  "result": {
    "element_count": 48130,
    "unexpected_count": 0,
    "unexpected_percent": 0.0,
    "partial_unexpected_list": [],
    "partial_unexpected_counts": [],
    "partial_unexpected_index_list": []
  },
  "meta": {},
  "exception_info": {
    "raised_exception": false,
    "exception_traceback": null,
```

```
      "exception_message": null
    }
}
```

That same expectation applied to the `cylinders` column will not be successful:

```
cylinders_exp = gx.expectations.ExpectColumnValuesToNotBeNull(
    column="cylinders"
)
result = batch.validate(cylinders_exp)
print(result)
```

```
{
  "success": false,
  "expectation_config": {
    "type": "expect_column_values_to_not_be_null",
    "kwargs": {
      "batch_id": "pandas_datasource-vehicles",
      "column": "cylinders"
    },
    "meta": {}
  },
  "result": {
    "element_count": 48130,
    "unexpected_count": 965,
    "unexpected_percent": 2.0049864949096197,
    "partial_unexpected_list": [
      null,
      null,
      ...
      null,
      null
    ],
    "partial_unexpected_counts": [
      {
        "value": null,
        "count": 20
      }
    ],
    "partial_unexpected_index_list": [
      7138,
      7139,
      8143,
```

```
      ...
      23022,
      23023,
      23024
    ]
  },
  "meta": {},
  "exception_info": {
    "raised_exception": false,
    "exception_traceback": null,
    "exception_message": null
  }
}
```

For brevity, we have only shown you how to set expectations around nullability, but there is an entire Expectations Gallery at `https://greatexpectations.io/expectations/` you can use for other assertions. Great Expectations also works with other tools like Spark, PostgreSQL, etc., so you can apply your expectations at many different points in your data transformation pipeline.

Visualization

Back in *Chapter 6, Visualization*, we discussed at length visualization using matplotlib, and we even discussed using Seaborn for advanced plots. These tools are great for generating static charts, but when you want to add some level of interactivity, you will need to opt for other libraries.

For this recipe, we are going to load the same data from the vehicles dataset we used back in our *Scatter plots* recipe from *Chapter 6*:

```
df = pd.read_csv(
    "data/vehicles.csv.zip",
    dtype_backend="numpy_nullable",
    dtype={
        "rangeA": pd.StringDtype(),
        "mfrCode": pd.StringDtype(),
        "c240Dscr": pd.StringDtype(),
        "c240bDscr": pd.StringDtype()
    }
)
```

```
df.head()
```

	barrels08	barrelsA08	charge120	...	phevCity	phevHwy	phevComb
0	14.167143	0.0	0.0	...	0	0	0
1	27.046364	0.0	0.0	...	0	0	0
2	11.018889	0.0	0.0	...	0	0	0

```
3   27.046364   0.0              0.0        ... 0              0              0
4   15.658421   0.0              0.0        ... 0              0              0
5 rows × 84 columns
```

Plotly

Let's start by looking at Plotly, which can be used to create visualizations with a high degree of inter-activity, making it a popular choice within Jupyter notebooks. To use it, simply pass `plotly` as the `backend=` argument to `pd.DataFrame.plot`. We are also going to add a `hover_data=` argument, which Plotly can use to add labels to each data point:

```
df.plot(
    kind="scatter",
    x="city08",
    y="highway08",
    backend="plotly",
    hover_data={"make": True, "model": True, "year": True},
)
```

If you inspect this in a Jupyter notebook or HTML page, you will see that you can hover over any data point to reveal more details:

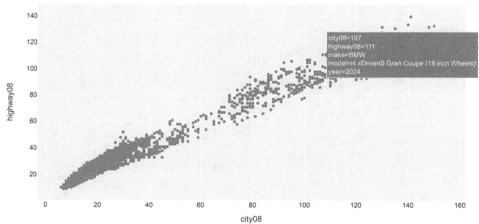

Figure 11.5: Hovering over a data point with Plotly

You can even select an area of the chart to zoom into the data points:

```
In [6]: df.plot(
            kind="scatter",
            x="city08",
            y="highway08",
            backend="plotly",
            hover_data={"make": True, "model": True, "year": True}
        )
```

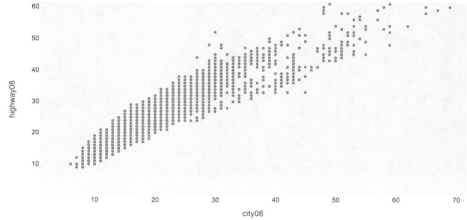

Figure 11.6: Zooming in with Plotly

As you can see, Plotly is very easy to use with the same pandas API you have seen throughout this book. If you desire interactivity with your plots, it is a great tool to make use of.

PyGWalker

All of the plotting code you have seen so far is declarative in nature; i.e., you tell pandas that you want a bar, line, scatter plot, etc., and pandas generates that for you. However, many users may prefer having a more "free-form" tool for exploration, where they can just drag and drop elements to make charts on the fly.

If that is what you are after, then you will want to take a look at the PyGWalker library. With a very succinct API, you can generate an interactive tool within a Jupyter notebook, with which you can drag and drop different elements to generate various charts:

```
import pygwalker as pyg
pyg.walk(df)
```

```
In [5]: import pygwalker as pyg
        pyg.walk(df)
```

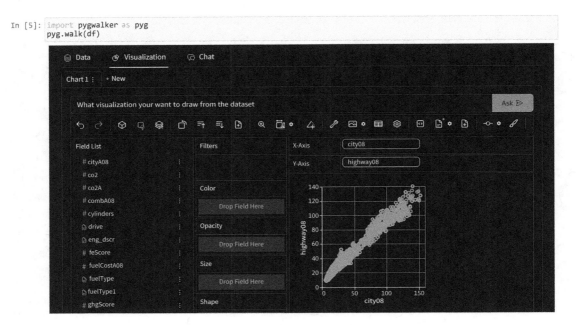

Figure 11.7: PyGWalker within a Jupyter notebook

Data science

While pandas offers some built-in statistical algorithms, it cannot hope to cover all of the statistical and machine learning algorithms that are used in the domain of data science. Fortunately, however, many of the libraries that do specialize further in data science offer very tight integrations with pandas, letting you move data from one library to the next rather seamlessly.

scikit-learn

scikit-learn is a popular machine learning library that can help with both supervised and unsupervised learning. The scikit-learn library offers an impressive array of algorithms for classification, prediction, and clustering tasks, while also providing tools to pre-process and cleanse your data.

We cannot hope to cover all of these features, but for the sake of showcasing something, let's once again load the vehicles dataset:

```
df = pd.read_csv(
    "data/vehicles.csv.zip",
    dtype_backend="numpy_nullable",
    dtype={
        "rangeA": pd.StringDtype(),
        "mfrCode": pd.StringDtype(),
        "c240Dscr": pd.StringDtype(),
        "c240bDscr": pd.StringDtype()
```

```
    }
)
```

```
df.head()
```

	barrels08	barrelsA08	charge120	...	phevCity	phevHwy	phevComb
0	14.167143	0.0	0.0	...	0	0	0
1	27.046364	0.0	0.0	...	0	0	0
2	11.018889	0.0	0.0	...	0	0	0
3	27.046364	0.0	0.0	...	0	0	0
4	15.658421	0.0	0.0	...	0	0	0

5 rows × 84 columns

Now let's assume that we want to create an algorithm to predict the combined mileage a vehicle will achieve, inferring it from other attributes in the data. Since mileage is a continuous variable, we can opt for a linear regression model to make our predictions.

The linear regression model we are going to work with will want to use features that are also numeric. While there are ways we could artificially convert some of our non-numeric data into numeric (e.g., using the technique from the *One-hot encoding with pd.get_dummies* recipe back in *Chapter 5*, *Algorithms and How to Apply Them*), we are just going to ignore any non-numeric columns for now. The linear regression model is also unable to handle missing data. We know from the *Exploring continuous data* recipe from *Chapter 6* that this dataset has two continuous variables with missing data. While we could try to interpolate those values, we are again going to take the simple route in this example and just drop them:

```
num_df = df.select_dtypes(include=["number"])
num_df = num_df.drop(columns=["cylinders", "displ"])
```

The scikit-learn model will need to know the *features* we want to use for prediction (commonly notated as X) and the target variable we are trying to predict (commonly notated as y). It is also a good practice to split the data into training and testing datasets, which we can do with the train_test_split function:

```
from sklearn.model_selection import train_test_split

target_col = "comb08"
X = num_df.drop(columns=[target_col])
y = num_df[target_col]

X_train, X_test, y_train, y_test = train_test_split(X, y)
```

With our data in this form, we can go ahead and train the linear regression model and then apply it to our test data to generate predictions:

```
from sklearn import linear_model
```

```
regr = linear_model.LinearRegression()
regr.fit(X_train, y_train)
y_pred = regr.predict(X_test)
```

Now that we have predictions from our test dataset, we can compare them back to the actual values we withheld as part of testing. This is a good way to measure how accurate the model is that we fit.

There are many different ways to manage model accuracy, but for now, we can opt for the commonly used and relatively simple mean_squared_error, which scikit-learn also provides as a convenience function:

```
from sklearn.metrics import import mean_squared_error


mean_squared_error(y_test, y_pred)
```

```
0.11414180317382835
```

If you are interested in knowing more, I highly recommend you read through the documentation and examples on the scikit-learn website, or check out books like *Machine Learning with PyTorch and Scikit-Learn: Develop machine learning and deep learning models with Python* (https://www.packtpub.com/en-us/product/machine-learning-with-pytorch-and-scikit-learn-9781801819312).

XGBoost

For another great machine learning library, let's now turn our attention to XGBoost, which implements algorithms using Gradient boosting. XGBoost is extremely performant, scales well, scores well in machine learning competitions, and pairs well with data that is stored in a pd.DataFrame. If you are already familiar with scikit-learn, the API it uses will feel familiar.

XGBoost can be used for both classification and regression. Since we just performed a regression analysis with scikit-learn, let's now work through a classification example where we try to predict the make of a vehicle from the numeric features in the dataset.

The vehicles dataset we are working with has 144 different makes. For our analysis, we are going to just pick a small subset of consumer brands:

```
brands = {"Dodge", "Toyota", "Volvo", "BMW", "Buick", "Audi", "Volkswagen",
"Subaru"}
df2 = df[df["make"].isin(brands)]
df2 = df2.drop(columns=["cylinders", "displ"])
```

From there, we are going to split our data into features (X) and a target variable (y). For the purposes of the machine learning algorithm, we also need to convert our target variable into categorical data type, so that the algorithm can predict values like 0, 1, 2, etc instead of "Dodge," "Toyota," "Volvo," etc.:

```
X = df2.select_dtypes(include=["number"])
y = df2["make"].astype(pd.CategoricalDtype())
```

With that out of the way, we can once again use the `train_test_split` function from scikit-learn to create training and testing data. Note that we are using `pd.Series.cat.codes` to use the numeric value assigned to our categorical data type, rather than the string:

```
X_train, X_test, y_train, y_test = train_test_split(X, y.cat.codes)
```

Finally, we can import the `XGBClassifier` from XGBoost, train it on our data, and apply it to our test features to generate predictions:

```
from xgboost import XGBClassifier

bst = XGBClassifier()
bst.fit(X_train, y_train)
preds = bst.predict(X_test)
```

Now that we have the predictions, we can validate how many of them matched the target variables included as part of our testing data:

```
accuracy = (preds == y_test).sum() / len(y_test)
print(f"Model prediction accuracy is: {accuracy:.2%}")
```

```
Model prediction accuracy is: 97.07%
```

Once again, we are only scratching the surface of what you can do with a library like XGBoost. There are many different ways to tweak your model to improve accuracy, prevent over-/underfitting, optimize for a different outcome, etc. For users wanting to learn more about this great library, I advise checking out the XGBoost documentation or books like *Hands-On Gradient Boosting with XGBoost and scikit-learn*.

Databases

Database knowledge is an important tool in the toolkit of any data practitioner. While pandas is a great tool for single-machine, in-memory computations, databases offer a very complementary set of analytical tools that can help with the storage and distribution of analytical processes.

Back in *Chapter 4, The pandas I/O System*, we walked through how to transfer data between pandas and theoretically any database. However, a relatively more recent database called DuckDB is worth some extra consideration, as it allows you to even more seamlessly bridge the worlds of dataframes and databases together.

DuckDB

DuckDB is a lightweight database system that offers a zero-copy integration with Apache Arrow, a technology that also underpins efficient data sharing and usage with pandas. It is extremely lightweight and, unlike most database systems, can be easily embedded into other tools or processes. Most importantly, DuckDB is optimized for analytical workloads.

DuckDB makes it easy to query data in your pd.DataFrame using SQL. Let's see this in action by loading the vehicles dataset:

```
df = pd.read_csv(
    "data/vehicles.csv.zip",
    dtype_backend="numpy_nullable",
    dtype={
        "rangeA": pd.StringDtype(),
        "mfrCode": pd.StringDtype(),
        "c240Dscr": pd.StringDtype(),
        "c240bDscr": pd.StringDtype()
    }
)

df.head()
```

```
     barrels08  barrelsA08  charge120  ...  phevCity  phevHwy  phevComb
0    14.167143  0.0         0.0        ...  0         0        0
1    27.046364  0.0         0.0        ...  0         0        0
2    11.018889  0.0         0.0        ...  0         0        0
3    27.046364  0.0         0.0        ...  0         0        0
4    15.658421  0.0         0.0        ...  0         0        0
5 rows × 84 columns
```

By passing a CREATE TABLE statement to duckdb.sql, you can load the data from the pd.DataFrame into a table:

```
import duckdb

duckdb.sql("CREATE TABLE vehicles AS SELECT * FROM df")
```

Once the table is created, you can query from it with SQL:

```
duckdb.sql("SELECT COUNT(*) FROM vehicles WHERE make = 'Honda'")
```

```
┌──────────────┐
│ count_star() │
│    int64     │
├──────────────┤
│         1197 │
└──────────────┘
```

If you want to convert your results back to a pd.DataFrame, you use the .df method:

```
duckdb.sql(
```

```
    "SELECT make, model, year, id, city08 FROM vehicles where make = 'Honda'
  LIMIT 5"
).df()
```

	make	model	year	id	city08
0	Honda	Accord Wagon	1993	10145	18
1	Honda	Accord Wagon	1993	10146	20
2	Honda	Civic Del Sol	1994	10631	25
3	Honda	Civic Del Sol	1994	10632	30
4	Honda	Civic Del Sol	1994	10633	23

For a deeper dive into DuckDB, I strongly advise checking out the DuckDB documentation and, for a greater understanding of where it fits in the grand scheme of databases, the *Why DuckDB* article (`https://duckdb.org/why_duckdb`). Generally, DuckDB's focus is on single-user analytics, but if you are interested in a shared, cloud-based data warehouse, you may also want to look at MotherDuck (`https://motherduck.com/`).

Other DataFrame libraries

Soon after pandas was developed, it became the de facto DataFrame library in the Python space. Since then, many new DataFrame libraries have been developed in the space, which all aim to address some of the shortcomings of pandas while introducing their own novel design decisions.

Ibis

Ibis is yet another amazing analytics tool created by Wes McKinney, the creator of pandas. At a high level, Ibis is a DataFrame "frontend" that gives you one generic API through which you can query multiple "backends."

To help understand what that means, it is worth contrasting that with the design approach of pandas. In pandas, the API or "frontend" for a group by and a sum looks like this:

```
df.groupby("column").agg(result="sum")
```

From this code snippet, the frontend of pandas defines how the query looks (i.e., for a group-by the operation you must call `pd.DataFrame.groupby`). Behind the scenes, pandas dictates how the `pd.DataFrame` is stored (in memory using pandas' own representation) and even dictates how the summation should be performed against that in-memory representation.

In Ibis, a similar expression would look like this:

```
df.group_by("column").agg(result=df.sum())
```

While the API exposed to the user may not be all that different, the similarities between Ibis and pandas stop there. Ibis does not dictate how you store the data you are querying; it can be stored in BigQuery, DuckDB, MySQL, PostgreSQL, etc., and it can be even stored in another DataFrame library like pandas. Beyond the storage, Ibis does not dictate how summation should be performed; instead, it leaves it to an execution engine. Many SQL databases have their own execution engine, but others may defer to third-party libraries like Apache DataFusion (`https://datafusion.apache.org/`).

To use a pd.DataFrame through Ibis, you will need to wrap it with the ibis.memtable function:

```
import ibis

df = pd.read_csv(
    "data/vehicles.csv.zip",
    dtype_backend="numpy_nullable",
    usecols=["id", "year", "make", "model", "city08"],
)

t = ibis.memtable(df)
```

With that out of the way, you can then start to query the data just as you would with pandas but using the Ibis API:

```
t.filter(t.make == "Honda").select("make", "model", "year", "city08")
```

```
r0 := InMemoryTable
  data:
    PandasDataFrameProxy:
            city08     id         make                model  year
        0       19      1   Alfa Romeo   Spider Veloce 2000  1985
        1        9     10      Ferrari            Testarossa  1985
        2       23    100        Dodge              Charger  1985
        3       10   1000        Dodge   B150/B250 Wagon 2WD  1985
        4       17  10000       Subaru     Legacy AWD Turbo  1993
        ...     ...    ...          ...                  ...   ...
    48125       19   9995       Subaru               Legacy  1993
    48126       20   9996       Subaru               Legacy  1993
    48127       18   9997       Subaru           Legacy AWD  1993
    48128       18   9998       Subaru           Legacy AWD  1993
    48129       16   9999       Subaru     Legacy AWD Turbo  1993

    [48130 rows x 5 columns]

  r1 := Filter[r0]
    r0.make == 'Honda'

Project[r1]
  make:   r1.make
  model:  r1.model
  year:   r1.year
  city08: r1.city08
```

It is worth noting that the preceding code does not actually return a result. Unlike pandas, which executes all of the operations you give it *eagerly*, Ibis collects all of the expressions you want and waits to perform execution until explicitly required. This practice is commonly called *deferred* or *lazy* execution.

The advantage of deferring is that Ibis can find ways to optimize the query that you are telling it to perform. Our query is asking Ibis to find all rows where the make is Honda and then select a few columns, but it might be faster for the underlying database to select the columns first and then perform the filter. How that works is abstracted from the end user; users are just required to tell Ibis what they want and Ibis takes care of how to retrieve that data.

To materialize this back into a `pd.DataFrame`, you can chain in a call to `.to_pandas`:

```
t.filter(t.make == "Honda").select("make", "model", "year", "city08").to_
pandas().head()
```

	make	model	year	city08
0	Honda	Accord Wagon	1993	18
1	Honda	Accord Wagon	1993	20
2	Honda	Civic Del Sol	1994	25
3	Honda	Civic Del Sol	1994	30
4	Honda	Civic Del Sol	1994	23

However, you are not required to return a `pd.DataFrame`. If you wanted a PyArrow table instead, you could opt for `.to_pyarrow`:

```
t.filter(t.make == "Honda").select("make", "model", "year", "city08").to_
pyarrow()
```

```
pyarrow.Table
make: string
model: string
year: int64
city08: int64
----
make: [["Honda","Honda","Honda","Honda","Honda",...,"Honda","Honda","Honda",
"Honda","Honda"]]
model: [["Accord Wagon","Accord Wagon","Civic Del Sol","Civic Del Sol",
"Civic Del Sol",...,"Prelude","Prelude","Prelude","Accord","Accord"]]
year: [[1993,1993,1994,1994,1994,...,1993,1993,1993,1993,1993]]
city08: [[18,20,25,30,23,...,21,19,19,19,21]]
```

For more information on Ibis, be sure to check out the Ibis documentation. There is even an Ibis tutorial aimed specifically at users coming from pandas.

Dask

Another popular library that has a history closely tied to pandas is Dask. Dask is a framework that provides a similar API to the `pd.DataFrame` but scales its usage to parallel computations and datasets that exceed the amount of memory available on your system.

If we wanted to convert our vehicles dataset to a Dask DataFrame, we can use the `dask.dataframe.from_pandas` function with a `npartitions=` argument that controls how to divide up the dataset:

```
import dask.dataframe as dd
ddf = dd.from_pandas(df, npartitions=10)
```

```
/home/willayd/clones/Pandas-Cookbook-Third-Edition/lib/python3.9/site-packages/
dask/dataframe/__init__.py:42: FutureWarning:
Dask dataframe query planning is disabled because dask-expr is not installed.

You can install it with `pip install dask[dataframe]` or `conda install dask`.
This will raise in a future version.

  warnings.warn(msg, FutureWarning)
```

By splitting your DataFrame into different partitions, Dask allows you to perform computations against each partition in parallel, which can help immensely with performance and scalability.

Much like Ibis, Dask performs calculations lazily. If you want to force a calculation, you will want to call the `.compute` method:

```
ddf.size.compute()
```

```
3991932
```

To go from a Dask DataFrame back to pandas, simply call `ddf.compute`:

```
ddf.compute().head()
```

```
     city08   id     make         model               year
0    19       1      Alfa Romeo   Spider Veloce 2000   1985
1    9        10     Ferrari      Testarossa           1985
2    23       100    Dodge        Charger              1985
3    10       1000   Dodge        B150/B250 Wagon 2WD  1985
4    17       10000  Subaru       Legacy AWD Turbo     1993
```

Polars

Polars is a newcomer to the DataFrame space and has developed impressive features and a dedicated following in a very short amount of time. The Polars library is Apache Arrow native, so it has a much cleaner type system and consistent missing value handling than what pandas offers today (for the history of the pandas type system and all of its flaws, be sure to give *Chapter 3, Data Types*, a good read).

In addition to a simpler and cleaner type system, Polars can scale to datasets that are larger than memory, and it even offers a lazy execution engine coupled with a query optimizer that can make it easier to write performant, scalable code.

For a naive conversion from pandas to Polars, you can use `polars.from_pandas`:

```python
import polars as pl

pl_df = pl.from_pandas(df)
pl_df.head()
```

```
shape: (5, 84)
barrels08   barrelsA08   charge120   charge240   ...   phevCity   phevHwy   phevComb
f64         f64          f64         f64         ...   i64        i64       i64
14.167143   0.0          0.0         0.0         ...   0          0         0
27.046364   0.0          0.0         0.0         ...   0          0         0
11.018889   0.0          0.0         0.0         ...   0          0         0
27.046364   0.0          0.0         0.0         ...   0          0         0
15.658421   0.0          0.0         0.0         ...   0          0         0
```

For lazy execution, you will want to try out the `pl.LazyFrame`, which can take the `pd.DataFrame` directly as an argument:

```python
lz_df = pl.LazyFrame(df)
```

Much like we saw with Ibis, the lazy execution engine of Polars can take care of optimizing the best path for doing a filter and select. To execute the plan, you will need to chain in a call to `pl.LazyFrame.collect`:

```python
lz_df.filter(
    pl.col("make") == "Honda"
).select(["make", "model", "year", "city08"]).collect().head()
```

```
shape: (5, 4)
make      model            year   city08
str       str              i64    i64
"Honda"   "Accord Wagon"   1993   18
"Honda"   "Accord Wagon"   1993   20
"Honda"   "Civic Del Sol"  1994   25
"Honda"   "Civic Del Sol"  1994   30
"Honda"   "Civic Del Sol"  1994   23
```

If you would like to convert back to pandas from Polars, both the `pl.DataFrame` and `pl.LazyFrame` offer a `.to_pandas` method:

```python
lz_df.filter(
    pl.col("make") == "Honda"
).select(["make", "model", "year", "city08"]).collect().to_pandas().head()
```

```
        make     model          year   city08
0       Honda    Accord Wagon   1993   18
1       Honda    Accord Wagon   1993   20
2       Honda    Civic Del Sol  1994   25
3       Honda    Civic Del Sol  1994   30
4       Honda    Civic Del Sol  1994   23
```

For a more detailed look at Polars and all of the great things it has to offer, I suggest checking out the *Polars Cookbook* (https://www.packtpub.com/en-us/product/polars-cookbook-9781805121152).

cuDF

If you have a Nvidia device and the CUDA toolkit available to you, you may also be interested in cuDF. In theory, cuDF is a "drop-in" replacement for pandas; as long as you have the right hardware and tooling, it will take your pandas expressions and run them on your GPU, simply by importing cuDF before pandas:

```
import cudf.pandas
cudf.pandas.install()

import pandas as pd
```

Given the power of modern GPUs compared to CPUs, this library can offer users a significant performance boost without having to change the way code is written. For the right users with the right hardware, that type of out-of-the-box performance boost can be invaluable.

packt.com

Subscribe to our online digital library for full access to over 7,000 books and videos, as well as industry leading tools to help you plan your personal development and advance your career. For more information, please visit our website.

Why subscribe?

- Spend less time learning and more time coding with practical eBooks and Videos from over 4,000 industry professionals
- Improve your learning with Skill Plans built especially for you
- Get a free eBook or video every month
- Fully searchable for easy access to vital information
- Copy and paste, print, and bookmark content

At www.packt.com, you can also read a collection of free technical articles, sign up for a range of free newsletters, and receive exclusive discounts and offers on Packt books and eBooks.

Other Books You May Enjoy

If you enjoyed this book, you may be interested in these other books by Packt:

Machine Learning with PyTorch and Scikit-Learn

Sebastian Raschka

Yuxi (Hayden) Liu

Vahid Mirjalili

ISBN: 978-1-80181-931-2

- Explore frameworks, models, and techniques for machines to learn from data
- Use scikit-learn for machine learning and PyTorch for deep learning
- Train machine learning classifiers on images, text, and more
- Build and train neural networks, transformers, and boosting algorithms
- Discover best practices for evaluating and tuning models
- Predict continuous target outcomes using regression analysis
- Dig deeper into textual and social media data using sentiment analysis

Deep Learning with TensorFlow and Keras, Third Edition

Amita Kapoor

Antonio Gulli

Sujit Pal

ISBN: 978-1-80323-291-1

- Learn how to use the popular GNNs with TensorFlow to carry out graph mining tasks
- Discover the world of transformers, from pretraining to fine-tuning to evaluating them
- Apply self-supervised learning to natural language processing, computer vision, and audio signal processing
- Combine probabilistic and deep learning models using TensorFlow Probability
- Train your models on the cloud and put TF to work in real environments
- Build machine learning and deep learning systems with TensorFlow 2.x and the Keras API

Machine Learning for Algorithmic Trading, Second Edition

Stefan Jansen

ISBN: 978-1-83921-771-5

- Leverage market, fundamental, and alternative text and image data
- Research and evaluate alpha factors using statistics, Alphalens, and SHAP values
- Implement machine learning techniques to solve investment and trading problems
- Backtest and evaluate trading strategies based on machine learning using Zipline and Backtrader
- Optimize portfolio risk and performance analysis using pandas, NumPy, and pyfolio
- Create a pairs trading strategy based on cointegration for US equities and ETFs
- Train a gradient boosting model to predict intraday returns using AlgoSeek s high-quality trades and quotes data

Packt is searching for authors like you

If you're interested in becoming an author for Packt, please visit authors.packtpub.com and apply today. We have worked with thousands of developers and tech professionals, just like you, to help them share their insight with the global tech community. You can make a general application, apply for a specific hot topic that we are recruiting an author for, or submit your own idea.

Join our community on Discord

Join our community's Discord space for discussions with the authors and other readers:

https://packt.link/pandas

Index

Symbols

.iloc
  item assignment with 38, 39
.loc
  item assignment with 38, 39
[] operator
  using, disadvantages 14, 15
[] operator, with DataFrame
  selecting 16, 17
  using, list argument 17
.pipe
  for chaining 149-152

A

advanced plots
  seaborn, using 214-223
aggregations 132-134
Apache Parquet
  using 105-112
apply 140-142
Arrow Database Connectivity (ADBC)
  using, with SQL 101-103

B

baseball players
  data, summarizing to show metrics 159, 160
  performance, comparing 290-298

position, finding that scored most runs for
  team over each season 161, 162
binary 124
binning algorithms 143-148
Boolean arrays
  filtering via 30-33
  selection via 30-33
Boolean types 50, 51

C

categorical data
  exploring 202-209
  type 56-59
chaining
  with .pipe 149-152
comma-separated values (CSV) files
  strategies, for reading 85-92
  using, for reading/writing 80-84
continuous data
  exploring 209-213
Copy on Write (CoW) 39
crime dataset
  calculating, by category 318-321
cuDF library 372

D

Dask library 370

databases
DuckDB 365-367

DataFrame
attributes 8-10
column assignment 39-43
joining, with pd.DataFrame.join 240, 241
label-based selection 24, 25
merging, with pd.merge 231-240
position-based selection 19, 20
working with 4, 5

DataFrame.filter 28, 29

DataFrame libraries 367
cuDF library 372
Dask 370
Ibis 367-369
Polars library 370-372

data mutation
avoiding 342

data science
scikit-learn 362-364
XGBoost 364, 365

data sizes
checking 338, 339

data type
selection by 29, 30

data validation
Great Expectations 356-359

DateOffsets 303-306

datetime 60-65
selecting 307-310

delimiter 80

dictionary-encode low cardinality data
working with 343

dtype=object
avoiding 334-338

DuckDB 365-367

E

exploratory data analysis
YData profiling 352-355

extract, load, transform (ELT) 75

extract, transform, load (ETL) 75

F

floating point types 48, 50

foundational libraries 350
NumPy 350, 351
PyArrow 351

G

Great Expectations
usage 356-359

group by
applying 275-278
usage 266-270

GroupBy concept 137

H

highest rated movies
selecting, by year 286-289

HTML 116-118

I

Ibis library 367-369

index 5-7

integral types 46-48

item assignment
with .loc and .iloc 38, 39

J

JavaScript Object Notation (JSON) 108-115

L

label-based selection
mixing, with position-based selection 25-28

lowest-budget movies
selecting, from top 100 152, 153

M

map 137, 138

masks 30
using 31-33

matplotlib
using, for plot customization 189-193

Microsoft Excel
hierarchical data 97, 98
tables, finding in non-default locations 95-97
used, for reading/writing files 93-95

missing values
handling 53-55

multiple columns
grouping and calculating 270-275

N

non-aggregated data
distribution, plotting of 183-188

NumPy 2, 350, 351

NumPy type system 72-77

O

object type 72-77

one-hot encoding
with pd.get_dummies 148, 149

P

pandas library 1
importing 2

pd.DataFrame
multiple columns, grouping
and calculating 270-275

pd.DataFrame arithmetic
basic structure 129-131

pd.DataFrame.assign method 42

pd.DataFrame.explode
reshaping with 257-260

pd.DataFrame.join
used, for joining DataFrames 240, 241

pd.DataFrame.melt
reshaping with 247-249

pd.DataFrame objects
concatenating 226-230

pd.DataFrame.pivot
reshaping with 252-256

pd.DataFrame.stack
reshaping with 242-246

pd.DataFrame.T
transposing 261-263

pd.DataFrame.unstack
reshaping with 242-246

pd.DatetimeIndex 309, 310

pd.get_dummies
one-hot encoding 148, 149

pd.merge
used, for merging DataFrames 231-240

pd.MultiIndex, with DataFrame
selecting with 37, 38

pd.MultiIndex, with multiple level
selecting with 34-36

pd.MultiIndex, with single level
selecting with 33, 34

pd.pivot_table
reshaping with 252-256

pd.RangeIndex 14

pd.Series
attributes 7, 8
creating 2, 3
label-based selection 21-23
position-based selection 17, 18
selecting 12-14

pd.Series arithmetic 126-128
basic structure 124
creating 124-126

pd.Series.plot
usage 167-169

pd.wide_to_long
reshaping with 250, 251

pickle format 119-121

plot customization
with matplotlib 189-193

Plotly 360, 361

Polars library 370-372

position-based selection
mixing, with label-based selection 25-28

PyArrow 351

PyArrow decimal types 70-72

PyArrow library 2

PyArrow List types 68, 69

PyGWalker 361

pytest library 348

R

reductions. See aggregations

resampling 310-315

Retrosheet
URL 159

S

scatter plots 193-202

scikit-learn 362-364

seaborn
using, for advanced plots 214-223

sensor-collected events
measuring, accurately with
missing values 322-332

SQL
ADBC, using with 101-104
SQLAlchemy, using with 99-101

SQLAlchemy
using, with SQL 99-101

stop order 155

string data type 51-53

summary statistics 142, 143

T

temporal PyArrow types 67, 68

temporal type
datetime 60-65
timedelta 65-67

test-driven development features 343-347

third-party I/O libraries 121

timedelta 65-67

timezone
handling 300-302

trailing stop order price
calculating 155-157

transformations 136, 137

tuple 3

U

unittest module 348

usage and performance tips
data mutation, avoiding 342
data sizes, checking 338
dictionary-encode low cardinality data 343

dtype=object, avoiding 333
test-driven development features 343
vectorized functions, using instead
    of loops 340

V

vectorized functions
using, instead of loops 340, 341
visualization 359
Plotly 360, 361
PyGWalker 361

W

weekly crime and traffic accidents
aggregating 315-317
window operations 278-285

X

XGBoost 364

Y

YData profiling 352-355

www.ingramcontent.com/pod-product-compliance
Lightning Source LLC
LaVergne TN
LVHW080111070326
832902LV00015B/2525

9781836205876